How to Pass

SECOND EDITION

NATIONAL 5

Biology

Billy Dickson and
Graham Moffat

HODDER GIBSON
AN HACHETTE UK COMPANY

Every effort has been made to trace all copyright holders, but if any have been inadvertently overlooked, the Publishers will be pleased to make the necessary arrangements at the first opportunity.

Although every effort has been made to ensure that website addresses are correct at time of going to press, Hodder Gibson cannot be held responsible for the content of any website mentioned in this book. It is sometimes possible to find a relocated web page by typing in the address of the home page for a website in the URL window of your browser.

Hachette UK's policy is to use papers that are natural, renewable and recyclable products and made from wood grown in well-managed forests and other controlled sources. The logging and manufacturing processes are expected to conform to the environmental regulations of the country of origin.

Orders: please contact Hachette UK Distribution, Hely Hutchinson Centre, Milton Road, Didcot, Oxfordshire, OX11 7HH. Telephone: +44 (0)1235 827827. Email education@hachette.co.uk. Lines are open from 9 a.m. to 5 p.m., Monday to Friday. You can also order through our website: www.hoddereducation.co.uk. If you have queries or questions that aren't about an order, you can contact us at hoddergibson@hodder.co.uk

First published in 2018 by
Hodder Gibson, an imprint of Hodder Education
An Hachette UK Company
50 Frederick Street
Edinburgh, EH2 1EX

Impression number   7

Year   2022

Cover photo © kikkerdirk/stock.adobe.com
Illustrations by Aptara, Inc.
Typeset in: Cronos Pro Light 13/15pt by Aptara, Inc.
Printed in India
A catalogue record for this title is available from the British Library.
ISBN: 9781510420830

We are an approved supplier on the Scotland Excel framework.

Find us on your school's procurement system as *Hachette UK Distribution Ltd* or *Hodder & Stoughton Limited t/a Hodder Education.*

Contents

Skills of Scientific Inquiry

Introduction

Welcome to *How to Pass National 5 Biology*!

The fact that you have opened this book and are reading it shows that you want to pass your SQA National 5 Biology course. This is excellent because passing, and passing well, needs that type of attitude. It also shows you are getting down to the revision that is essential to pass and get the best grade possible.

The idea behind this book is to help you to pass, and if you are already on track to pass, it can help you improve your grade. It can help boost a C into a B or a B into an A. It cannot do the work for you, but it can guide you in how best to use your limited time.

In producing this book we have assumed that you are following or have followed an SQA National 5 level Biology course at school or college this year and that you have probably, but not necessarily, worked on Science Experiences and Outcomes up to and including Fourth Level at an earlier stage.

We recommend that you download a copy of the National 5 level Biology Course Specification from the SQA website at www.sqa.org.uk.

Course assessment outline

The components of the National 5 Biology assessment system are summarised in the table below.

National 5 Biology	Assessment	Who does the assessing?
Course (graded A–D)	**1** Examination (worth 80% of the grade) 25 multiple-choice questions worth 1 mark each and 75 marks from structured and extended response questions	Marked by SQA out of 100 marks
	2 Assignment (worth 20% of the grade)	Marked by SQA out of 20 marks

Component 1: course examination (100 marks)

The National 5 examination is a single paper consisting of a booklet of questions in two sections:

- **Section 1** contains 25 multiple-choice questions for 1 mark each.
- **Section 2** contains a mixture of structured and extended response questions for a total of 75 marks. Some questions might have choice within them.

The majority of the marks test demonstration and application of knowledge, with an emphasis on the application of knowledge. The remainder test the application of scientific inquiry skills. Each paper includes a short passage testing scientific literacy.

The course examination is marked by SQA and contributes 80% to the overall grade for the course. A list of the knowledge and skills required for your exam is summarised in the table below.

Knowledge and skills	Meaning
Demonstrating knowledge	Demonstrating knowledge and understanding of key areas of biology
Applying knowledge	Applying knowledge of key areas of biology to new situations, interpreting information and solving problems
Planning	Planning or designing experiments to test given hypotheses
Selecting	Selecting information from a variety of sources
Presenting	Presenting information appropriately in a variety of forms
Processing	Processing information using calculations and units where appropriate
Predicting	Making predictions and generalisations based on evidence
Concluding	Drawing valid conclusions and giving explanations supported by evidence
Evaluating	Suggesting improvements to experiments and investigations

The knowledge content is detailed in the tables on pages 25–47 of the Course Specification and you should note that both the Key Area and Depth of Knowledge columns can be examined. You will also be expected to be familiar with the apparatus and techniques in the table on page 48.

It is a good idea to get copies of any specimen or past papers that are available on the SQA website.

Component 2: Assignment (20 marks)

The assignment is a task which is based on experiments and further research that you have carried out in class time. The investigation will be supervised by your teacher and you will have to write up the work in the form of a report during a controlled assessment. During the controlled assessment you will have access to selected material and notes but you cannot use a draft copy of any part of your assignment report.

The assignment has two stages:
1 a research stage
2 a report stage.

The assignment report

The report is marked out of 20 marks, allocated as shown in the table below. There are 17 marks for skills and 3 marks for the demonstration and application of biological knowledge.

Aspect of your report	Marks
Stating your aim	1
Underlying biology	3
Data collection and handling	6
Graphical presentation	4
Analysis	1
Drawing a valid conclusion	1
Evaluation	2
Structuring your report	2
Total	**20**

The assignment is marked by SQA and makes up 20% of your overall course assessment and grading.

About this book

We have tried to keep this book simple and easy to understand and have used the language of the SQA National 5 Course Specification. This is the language used in the setting of the exam papers.

We suggest that you use this book throughout your course. Use it at the end of each Key Area covered in class, in preparation for any assessment or tests, before your preliminary examination and, finally, to revise the whole course in the lead-up to your final examination.

There is a grid on page x that you can use to record and evaluate your progress.

The course content is split into three sections which cover the three areas of National 5 Biology. Each section is divided into chapters, each of which covers a Key Area. The Key Areas have four main features: key points, summary notes, key words and questions.

Key points

This is a list of the pieces of knowledge you will need to demonstrate and apply in each Key Area. Each point ends with a tick box which you can use to monitor your learning. You could traffic-light each tick box: green for fully understood, orange for nearly there and red for needing to be looked at again. The words in **bold** within the key points are vital and each is defined in the key words section at the end of the chapter and also in the glossary at the back of the book.

Summary notes

These give a summary of the knowledge required in each Key Area. You must read these carefully. You could use a highlighter pen to emphasise certain words or phrases and you might want to add your own notes in the margin in pencil or on sticky notes. There are diagrams to illustrate many of the key learning ideas. The summary notes often contain Hints & tips boxes. In addition, the symbol shown on the right indicates that a technique which you need to be familiar with for your exam is described within the notes.

Key words

The key words used are listed at the end of each chapter along with a definition of each. These are the crucial terms you need to know to do well in N5 Biology. After completing your work on a Key Area, it is a good idea to make a set of flash cards using these words.

Questions

A set of both short-answer and longer-answer questions is found at the end of each chapter. These allow you to practise the demonstration of knowledge from each Key Area. National standard answers are provided at the end of each set of chapters and it is recommended that you mark and correct your own work.

Practice assessment

We have included a practice assessment linked to each area. These can give you an idea of your overall progress in the course. We have designed these assessments to be like mini course exams with multiple-choice, structured response and extended response items.

The questions are intended to replicate the style of those you will meet in the course exam. They allow you to judge how you are doing overall. Some questions will test knowledge and its application while others will test your skills of scientific inquiry; these have been indicated with 'SSI' so that you can identify them. These are provided in roughly the same proportion as in your final exam.

Give yourself a maximum of 70 minutes to complete each test.

Mark your own work using the answers provided at the end of each section. You could grade your work as you go along to give you an idea of how well you are doing in the course. The table below shows a suggested grading system:

Mark out of 45	Grade
18–22 marks	D
23–26 marks	C
27–31 marks	B
32+ marks	A

Skills of scientific inquiry

This section offers three different approaches to revising and improving your skills of scientific inquiry. Section 4.1 focuses on an example investigation and provides questions on the thinking that should go into experimental design, collection and presentation of results and concluding from these results. Section 4.2 involves seven practice questions in which all the individual skills, apparatus and techniques have been identified for you so that you can work on consolidating your strengths and improving weaker areas. In Section 4.3, we offer some classified tips and hints for tackling exam questions. Most students should use all three sections. Answers and a set of key words are provided at the end of the section.

Your assignment

We give an introduction to the assignment, some suggestions for suitable topics and some information to help you complete the task. The table on page 150 has a checklist of points you will need to have in your report.

Your exam

We give some hints on approaches to your final exams in general as well as more specific tips for your National 5 Biology exam.

Glossary

Here we have brought together the meanings of the key words that occur in the Course Specification for National 5 Biology in the context of the Key Areas in which they first appear in the book. You could use this glossary to make flash cards. A flash card has the term on one side and the definition on the other. Get together with a friend and use these cards to test each other.

Record of progress and self-evaluation checklist

Use the grid below to record and evaluate your progress as you finish each of the Key Areas and try the practice assessments.

Key Area			Progress indicator				Key Area feedback 1: OK 2: more work 3: help needed
			Key points... Traffic-lighted	Summary notes... Read and understood	Questions... Tried and self-marked	Key words... Flash cards made	
Cell Biology	1.1	Cell structure					
	1.2	Transport across cell membranes					
	1.3	DNA and the production of proteins					
	1.4	Proteins					
	1.5	Genetic engineering					
	1.6	Respiration					
	Practice course assessment		**Section 1/10**		**Section 2/35**		**Total/45**
Multicellular Organisms	2.1	Producing new cells					
	2.2a	Control and communication: nervous control					
	2.2b	Control and communication: hormonal control					
	2.3	Reproduction					
	2.4	Variation and inheritance					
	2.5	Transport systems: plants					
	2.6	Transport systems: animals					
	2.7	Absorption of materials					
	Practice course assessment		**Section 1/10**		**Section 2/35**		**Total/45**
Life on Earth	3.1	Ecosystems					
	3.2	Distribution of organisms					
	3.3	Photosynthesis					
	3.4	Energy in ecosystems					
	3.5	Food production					
	3.6	Evolution of species					
	Practice course assessment		**Section 1/10**		**Section 2/35**		**Total/45**

Cell Biology

Cell structure

Key points !

1. A **unicellular** organism has one single **cell**. ☐
2. A **multicellular** organism is made up of many cells. ☐
3. The **ultrastructure** of a cell is its fine structure as revealed at high magnification. ☐
4. The ultrastructure of an animal cell includes its **cell membrane, nucleus, cytoplasm, mitochondria** and **ribosomes**. ☐
5. The ultrastructure of a plant cell includes its **cell wall**, cell membrane, nucleus, cytoplasm, sap **vacuole**, mitochondria and ribosomes. ☐
6. **Chloroplasts** are present in green plant cells. ☐
7. The ultrastructure of a **fungal cell** includes its cell wall, cell membrane, nucleus, cytoplasm, vacuole, mitochondria and ribosomes. ☐
8. The ultrastructure of a **bacterial cell** includes its cell wall, cell membrane, cytoplasm, **plasmids** and ribosomes. ☐
9. The nucleus contains genetic material (**DNA**) and controls cell activities. ☐
10. The cell membrane controls the entry and exit of substances into and out of the cell. ☐
11. Cell **organelles**, such as nuclei, mitochondria and chloroplasts, are compartments found in the cytoplasm and are the site of chemical reactions. ☐
12. Bacterial cells do not contain organelles. ☐
13. The cell wall is for support and shape and prevents plant cells from bursting. ☐
14. Plant cell walls are made of **cellulose** whereas those of fungal and bacterial cells are made of different materials and have different structure. ☐
15. The vacuole stores a watery solution of salts and sugars and helps to support cells. ☐
16. Chloroplasts contain chlorophyll and are the sites of **photosynthesis**. ☐
17. Mitochondria are the sites of **aerobic respiration** in cells. ☐
18. Ribosomes are the sites of **protein** synthesis in cells. ☐
19. In bacterial cells, plasmids hold some of the genetic material (DNA) of the cell. ☐
20. Cells can be magnified for study using a **microscope**. ☐
21. Bacterial cells can be cultured for study in shallow transparent containers called **Petri dishes**. ☐

Summary notes

Living organisms and cells

The bodies of living organisms are made up of basic units called cells. Living organisms are either unicellular, with only one cell, or multicellular, with more than one cell.

Organisms can be divided into groups. These include animals, plants, fungi and bacteria.

Cell types

Animal cells are surrounded by a cell membrane and contain a nucleus, ribosomes and mitochondria.

Plant cells are usually bigger than other cells and are surrounded by a cellulose cell wall with a membrane inside. They contain a nucleus and cytoplasm with a large central vacuole, ribosomes and mitochondria. Green plant cells also have chloroplasts.

Fungal cells are surrounded by a cell wall with a membrane inside and contain a vacuole, a nucleus, ribosomes and mitochondria.

Bacterial cells are usually very much smaller than other cells and are surrounded by a cell wall with a membrane inside. Their cytoplasm contains plasmids and ribosomes, but no organelles. Bacteria can be cultured in the laboratory. Pure cultures of a single species of bacteria can be grown on nutrient-rich jelly in shallow transparent containers called Petri dishes.

Cell structure

The fine structure of cells, which can only be seen using a high-magnification microscope, is called ultrastructure. Ultrastructure includes organelles such as nuclei, mitochondria and chloroplasts. Other ultrastructure features such as ribosomes and plasmids are not organelles.

Some structures, such as ribosomes, are found in cells from all types of organism. Figure 1.1 shows the general structure of different cells.

> **Hints & tips** ⭐
>
> *For your exam it could be handy to remember that bacterial cells do not contain organelles. Bacterial cells are usually very much smaller than other cells.*

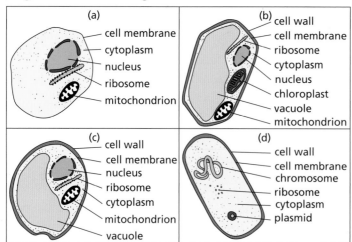

Figure 1.1 General structure of different cells (not to scale): **(a)** typical animal cell – a human cheek cell; **(b)** typical plant cell – a leaf mesophyll cell; **(c)** typical fungal cell – a yeast cell; **(d)** typical bacterial cell – a bacillus

Functions of cell structures

The different structures present in cells have different functions in the life of the cell. The different functions are shown in the table below and some of these are covered in other Key Areas, as indicated.

Cell structure	Function	Animal cell	Plant cell	Fungal cell	Bacterial cell
Nucleus	Contains genetic information (DNA) in animal, plant and fungal cells and so controls cell activities (see Key Areas 1.3 and 1.4 on pages 11 and 14)	✓	✓	✓	–
Plasmid	A small ring of genetic material in a bacterial cell	–	–	–	✓
Cell membrane	A selectively permeable membrane that controls entry and exit of substances such as O_2, CO_2, glucose and waste to and from all cells (see Key Area 1.2 on page 5)	✓	✓	✓	✓
Cytoplasm	Watery, jelly-like material within cells containing organelles that are the sites of various chemical reactions	✓	✓	✓	✓
Cell wall	The outer layer of plant, fungal and bacterial cells, which helps support the cell	–	✓	✓	✓
Vacuole	Membrane-bound sac that stores a solution of water, salts and sugars and helps support plant and fungal cells	–	✓	✓	–
Chloroplast	Makes carbohydrate in green plant cells using light energy in the process of photosynthesis (see Key Area 3.3 on page 99)	–	✓	–	–
Mitochondrion	Main site of ATP production in aerobic respiration in animal, plant and fungal cells (see Key Area 1.6 on page 21)	✓	✓	✓	–
Ribosome	Site of protein synthesis in cells (see Key Area 1.3 on page 11)	✓	✓	✓	✓

Note on cell walls

Cell walls are found in plant, fungal and bacterial cells. The cell walls from these different groups of living organism differ in their structure and in the chemical substances of which they are made. Plant cell walls are made of cellulose, whereas those of fungi and bacteria are varied in structure and chemical composition and do not contain cellulose.

> **Hints & tips** ⭐
>
> *The cell walls of plant cells are made of cellulose but those of fungi and bacteria are made of different materials.*

Key words

Aerobic respiration – release of energy from food by a cell using oxygen

Bacterial cell – the tiny individual cell of a bacterium

Cell – the basic unit of life

Cell membrane – selectively permeable membrane enclosing the cell cytoplasm and controlling the entry and exit of materials

Cell wall – supports and prevents cells from bursting; plant, fungal and bacterial walls have different structures and chemical compositions

Cellulose – structural carbohydrate of which plant cell walls are made

Chloroplast – organelle containing chlorophyll; the site of photosynthesis

Cytoplasm – jelly-like liquid which contains cell organelles and is the site of many chemical reactions

DNA – deoxyribonucleic acid; substance in chromosomes that carries the genetic code of an organism

Fungal cell – individual cell of a fungus

Microscope – an instrument for magnifying objects such as cells for study

Mitochondrion – organelle that is the site of aerobic respiration and ATP production in cells (*pl.* mitochondria)

Multicellular – having many cells

Nucleus – organelle which controls cell activities and contains the genetic information of the organism (*pl.* nuclei)

Organelle – membrane-bound compartment with a specific function in animal, plant and fungal cells

Petri dish – shallow transparent container used for growing bacteria and carrying out other experiments

Photosynthesis – process carried out by green plants to make their own food using light energy

Plasmid – circular genetic material present in bacterial cells and used in genetic engineering or modification

Protein – substance composed of chains of amino acids and containing the elements carbon, hydrogen, oxygen and nitrogen

Ribosome – site of protein synthesis

Ultrastructure – fine structure and detail of a cell and its organelles revealed by an electron microscope

Unicellular – single-celled

Vacuole – membrane-bound sac containing cell sap in plant and fungal cells

Questions ?

A Short-answer questions

1. Name **two** structures that are found in all cells. (2)
2. Give **two** organelles found in animal, plant and fungal cells. (2)
3. Name **one** structure found in bacterial cells only. (1)
4. Describe how the cell walls of plant, fungal and bacterial cells are different. (1)
5. Give the function of mitochondria. (1)
6. Give the function of a chloroplast. (1)
7. Give the function of a ribosome. (1)
8. Describe the following cell structures:
   a) a plant cell vacuole
   b) a bacterial cell plasmid. (2)

B Longer-answer questions

1. Describe the similarities and differences between an animal cell and a bacterial cell. (5)
2. Describe the structure and function of plant cell walls. (2)
3. Describe the similarities and differences in structure between a green plant cell and a bacterial cell. (5)
4. Give **one** difference and **one** similarity between the structure of a fungal and a bacterial cell. (2)

Answers on page 25

Key Area 1.2
Transport across cell membranes

Key points ❗

1 The cell membrane is composed of **phospholipid** and protein. ☐

2 The cell membrane is **selectively permeable**. ☐

3 A difference in concentration of a substance is a **concentration gradient**. ☐

4 **Passive transport** of a substance does not require additional energy and describes the movement of its molecules down a concentration gradient from a higher concentration to a lower concentration. ☐

5 **Diffusion** and **osmosis** are examples of passive transport processes. ☐

6 Diffusion is the movement of substances down a concentration gradient from a higher concentration to a lower concentration. ☐

7 Examples of substances that enter most cells by diffusion are oxygen, glucose and amino acids. ☐

8 An example of a substance that leaves most cells by diffusion is carbon dioxide. ☐

9 Diffusion is important to cells because it helps provide the cell with raw materials and helps to remove waste products. ☐

10 Osmosis is the movement of water molecules from a region of higher water concentration to a region of lower water concentration through a selectively permeable membrane. ☐

11 An animal cell placed in a solution with a water concentration higher than that inside the cell will take up water by osmosis and could burst. ☐

12 An animal cell placed in a solution with a lower water concentration than that inside the cell will lose water by osmosis and could shrink. ☐

13 A plant cell placed in a solution with a water concentration higher than that inside the cell will take up water by osmosis and become **turgid**. ☐

14 Turgid is the term used to describe a cell or tissue in which the vacuole has swollen due to water gain and presses the cytoplasm and cell membrane against the cell wall. ☐

15 A plant cell placed in a solution with a lower water concentration than that inside the cell will lose water by osmosis and become **plasmolysed**. ☐

16 Plasmolysed is the term used to describe a cell in which the vacuole has shrunk due to water loss, causing the cell membrane to pull away from the cell wall. ☐

17 **Active transport** is the movement of molecules from a region of low concentration to a region of higher concentration, against the concentration gradient. ☐

18 Active transport requires additional energy provided by ATP to allow membrane proteins to move molecules and ions against the concentration gradient. ☐

Summary notes
Cell membrane structure

Cell membranes are at, or just inside, the boundaries of all cells. Membranes are extremely thin and are composed of phospholipid and protein molecules in a layered arrangement containing pores (Figure 1.2).

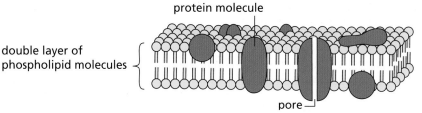

Figure 1.2 Structure of the cell membrane

Transport into and out of cells

The cells of a living organism exchange substances with each other and with their surroundings. This can happen in several ways including diffusion, osmosis and active transport. These terms all refer to the movement of molecules and involve concentration gradients, which are differences in concentration of these molecules between regions.

Diffusion and osmosis are passive because they do not require input of additional energy but active transport is active because of the additional energy it requires.

Hints & tips

There is more about the functions of proteins in Key Area 1.4 on page 14.

Diffusion

Diffusion is the passive movement of molecules down a concentration gradient from a region of higher concentration to a region of lower concentration. It continues until the molecules are spread out evenly. In living cells diffusion often occurs through membranes (Figure 1.3).

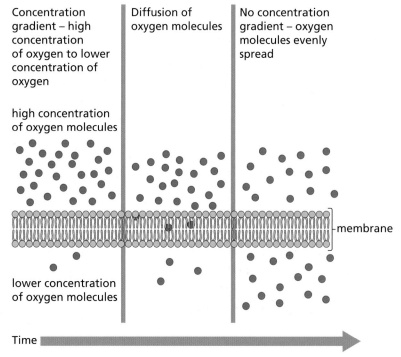

Figure 1.3 Diffusion of oxygen molecules through a cell membrane

A good example of diffusion in action is the exchange of gases in the lungs of a mammal. Oxygen molecules move from a region of higher concentration in the lungs to a region of lower concentration in the red blood cells (Figure 1.4). Carbon dioxide moves by diffusion in the opposite direction. Diffusion is also the method of absorption of glucose and amino acids through the small intestine of a mammal.

Hints & tips

There is more about gas exchange in the lungs and absorption of nutrients in the small intestine in Key Area 2.7 on page 72.

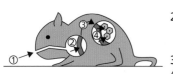

1 Breathes air with high concentration of oxygen molecules into lung spaces
2 Oxygen molecules diffuse from high concentration in lung spaces to lower concentration in the blood
3 Blood transported to cells of body
4 Oxygen molecules diffuse from high concentration in blood to lower concentration in body cells

Figure 1.4 Diffusion of oxygen in the body of a mammal

Hints & tips

Glucose and Oxygen enter cells by diffusion where they are involved in Respiration to release Energy.
Remember: GORE!

Osmosis

Osmosis is a special kind of diffusion. It is the passive movement of water molecules down a concentration gradient from a region of higher water concentration to a region of lower water concentration. Water movement by osmosis is always through a selectively permeable membrane (Figure 1.5).

Hints & tips

There is more about respiration in Key Area 1.6 on page 21.

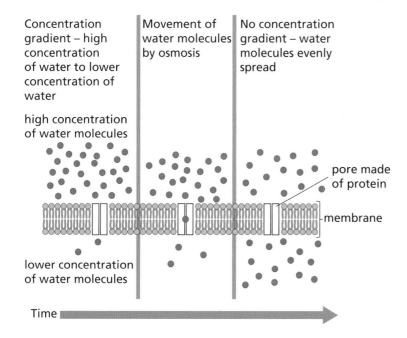

Figure 1.5 Osmosis

A good example of osmosis in action is the uptake of water into the root cells of plants. When the soil is moist after rain, the soil water concentration is higher than that of plant root cells. Water therefore passes from higher concentration in the soil to lower concentration in the root cells by osmosis.

Hints & tips

There is more about osmosis in plant roots in Key Area 2.5 on page 60.

Active transport

Active transport is the movement of molecules or ions from a region of lower concentration to a region of higher concentration. Active transport requires energy for membrane proteins to move molecules or ions against the concentration gradient.

In some cases it is essential that cells with a high concentration of a substance receive more of that substance from areas where the substance is at a lower concentration. A good example of this is found in certain unicellular freshwater plants such as *Nitella*. The cells of this plant require potassium ions from their environment to survive. However, potassium ion concentrations in freshwater are very low. *Nitella* cells use energy from their respiration process to take up potassium ions from low concentrations in freshwater to the higher concentrations that must be maintained in their cells. The energy goes to the cell membrane proteins, which carry the ions through the membranes from outside to inside (Figure 1.6).

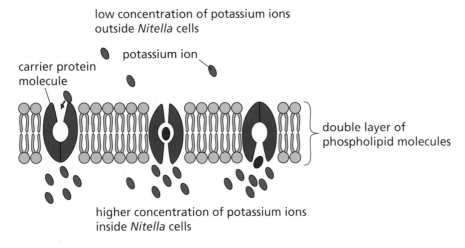

Figure 1.6 Active transport of potassium ions in *Nitella*

Animal cells and water

The results of placing animal cells in solutions of different water concentration can be shown using red blood cells (Figure 1.7). If animal cells are placed in a solution with a water concentration higher than the cell contents, water is taken in by osmosis and the cells will burst. If placed in a solution of lower water concentration than the cell contents, water will be lost by osmosis and the cell will shrink.

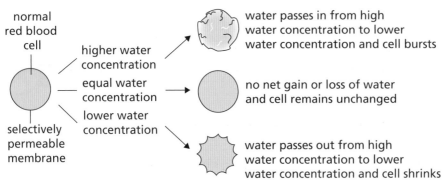

Figure 1.7 Effects of water on a red blood cell

Plant cells and water

The results of placing plant cells in solutions of different water concentration are shown in Figure 1.8. If plant cells are immersed in a solution with a water concentration higher than the cell contents, they will take up water by osmosis and their vacuoles will fill. Further intake of water is prevented because vacuole size is limited by the presence of the cell wall. Plant cells that have full vacuoles are said to be turgid; this is the normal state for a plant cell. Plant cells immersed in a solution with a lower water concentration than the cell contents lose water by osmosis and their vacuoles shrink and eventually pull the cell membrane and cytoplasm away from the cell walls of the cell. This is not desirable for the plant and may actually kill it. Cells with membranes pulled away from their walls are said to be plasmolysed.

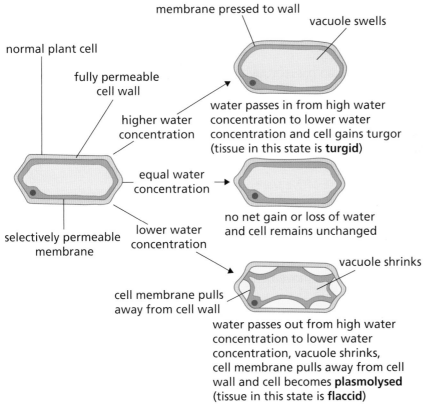

Figure 1.8 Effects of water on a plant cell

Key words

Active transport – transport of molecules against their concentration gradient
Concentration gradient – difference in concentration between two solutions, cells or solutions and cells
Diffusion – passive movement of molecules from an area of high concentration to an area of lower concentration
Osmosis – movement of water molecules from an area of high water concentration to an area of lower water concentration through a selectively permeable membrane
Passive transport – movement of molecules down a concentration gradient without the need for additional energy, e.g. diffusion and osmosis

Phospholipid – fat or oil with molecules composed of fatty acids and glycerol

Plasmolysed – description of a plant cell in which the vacuole has shrunk and the membrane has pulled away from the wall due to water loss

Selectively permeable – refers to a membrane that controls the movement of certain molecules depending on their size

Turgid – description of a swollen plant cell with a full vacuole resulting from water intake due to osmosis

Questions ?

A Short-answer questions

1 Name **two** substances present in cell membranes. (2)
2 State a feature of the cell membrane which allows the movement of only certain substances into the cell. (1)
3 Describe what is meant by a concentration gradient. (1)
4 Give **one** passive process by which substances can pass through membranes. (1)
5 Describe the process of diffusion. (2)
6 State the name given to the movement of water from an area of high water concentration to an area of lower water concentration through a selectively permeable membrane. (1)
7 State **two** differences between osmosis and diffusion. (2)
8 Describe the process of active transport. (2)
9 Give **one** example of a situation in which cells use active transport. (1)
10 State **one** difference between active and passive transport. (1)
11 Give the meanings of the following terms:
   a) turgid
   b) plasmolysed. (2)
12 In terms of water concentration, explain how water movements account for a plant cell becoming turgid. (1)
13 In terms of water concentration, explain how water movements account for a plant cell becoming plasmolysed. (1)
14 Describe **two** features of a plasmolysed plant cell. (2)

B Longer-answer questions

1 Explain the importance of diffusion in multicellular organisms. (3)
2 Describe the role of protein molecules in osmosis and in active transport. (2)
3 Explain what would happen to a piece of potato tissue placed into a beaker of pure water for 2 hours. (3)
4 Explain why immersing a single animal cell in sea water might kill it. (3)

Answers on page 25

Key Area 1.3
DNA and the production of proteins

Key points !

1 The nucleus of living cells contains genetic information organised into chromosomes. ☐
2 Chromosomes are made up of sections called **genes**. ☐
3 DNA is a complex substance that forms the genes of all living organisms. ☐
4 DNA carries the genetic information, which is information for making proteins. ☐
5 A single DNA molecule is a **double-stranded helix**. ☐
6 Each strand of DNA is a chain carrying molecules called **bases**. ☐
7 There are four different types of base known as **adenine (A), thymine (T), guanine (G)** and **cytosine (C)**. ☐
8 The strands of the DNA helix are held by **bonds** between bases on each strand. ☐
9 The bases bond together to form **complementary** base pairs. ☐
10 Base A only pairs with T and base G only pairs with C. ☐
11 The sequence of bases A, T, G and C make up the **genetic code**. ☐
12 The base sequence of a specific gene determines the **amino acid** sequence in the specific protein to be assembled. ☐
13 Proteins are assembled at ribosomes in the cell cytoplasm. ☐
14 **Messenger RNA (mRNA)** is a molecule that carries a complementary copy of the code from the DNA in the nucleus to a ribosome in the cytoplasm where the protein is assembled from amino acids. ☐

Summary notes
Structure of DNA

Genes are made from deoxyribonucleic acid (DNA) molecules. DNA is a substance with large, complex molecules. These molecules carry the genetic information of a cell coded into them. Each DNA molecule is in the shape of a double-stranded helix, as shown in Figure 1.9. Each strand carries chemical units called bases of which there are four different types. The bases are known by the upper-case first letters of their chemical names: **A**denine, **T**hymine, **G**uanine and **C**ytosine.

The two strands of the DNA molecule are held together by bonds between bases on each strand. A on one strand always pairs with T on the other strand; G on one strand always pairs with C on the other strand. These bases are said to be complementary to each other: A with T and G with C. Figure 1.10 shows details of the complementary base pairing in DNA.

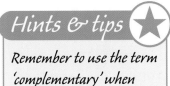

Hints & tips ★

Remember to use the term 'complementary' when talking about the base pairs. It means a perfect match.

Genetic code

The genetic code of a cell allows it to make the specific proteins it needs. Each gene has a code for a different protein. Different species have different genes and so are able to make different proteins.

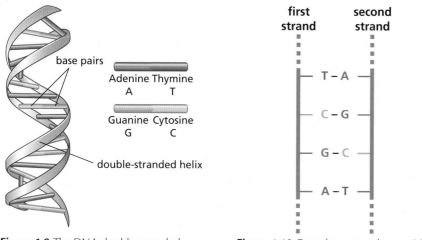

Figure 1.9 The DNA double-stranded helix – note how the two strands are linked by pairs of bases

Figure 1.10 Complementary base pairing

Each type of protein is made up from a chain of amino acids in an order determined by the genetic code. It is the sequence of bases along one strand of a DNA molecule that encodes its genetic information.

There are 20 different amino acids that occur naturally. Organisms obtain amino acids from their diet and transport them to each of their cells.

Protein synthesis

DNA is found in the nucleus of a cell while proteins are assembled from amino acids at the ribosomes found in the cytoplasm. A complementary copy of the code from the DNA in the nucleus is carried to a ribosome by a messenger RNA (mRNA) molecule. mRNA has the base **U**racil (U) in place of **T**hymine. Figure 1.11 shows a DNA code being copied as a complementary copy to mRNA and then carried to a ribosome to be translated into a protein molecule.

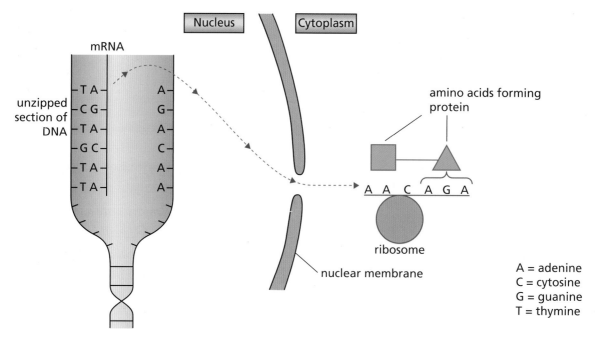

A = adenine
C = cytosine
G = guanine
T = thymine

Figure 1.11 Protein synthesis

Key words

A, G, T, C – letters that represent the names of bases of DNA and mRNA (adenine, guanine, thymine, cytosine)

Amino acid – a building block of a protein molecule

Bases – form the genetic code of DNA and mRNA

Bond – chemical link between atoms in a molecule

Complementary – fitting together as applied to DNA base pairing

Double-stranded helix – describes the spiral ladder shape of DNA molecules

Gene – small section of DNA that codes for the production of a specific protein

Genetic code – code formed by the sequence of the bases in DNA that determines an organism's characteristics

Messenger RNA (mRNA) – substance that carries a complementary copy of the genetic code from DNA to the ribosomes

Questions ?

A Short-answer questions

1   State the role of DNA in cells.   (1)
2   State what is meant by the term 'genetic information'.   (1)
3   Describe the overall shape of a DNA molecule.   (1)
4   Give the letters by which the four bases that make up the genetic code in a DNA molecule are known. (1)
5   Give the pairs of letters of the complementary bases in DNA.   (2)
6   Give the term that refers to a section of DNA with a specific sequence of bases that code for the production of a specific protein.   (1)
7   Name the molecule that carries a complementary copy of the base sequence of DNA out of the nucleus.   (1)
8   Name the basic units which are joined together to make a protein at the ribosome.   (1)
9   State the location within a cell where amino acids are assembled to produce protein molecules.   (1)

B Longer-answer questions

1   Describe the structure of DNA.   (3)
2   Explain the term 'complementary' in relation to base pairs in a molecule of DNA.   (2)
3   Describe the feature of an mRNA molecule that ensures the order of the amino acids in a protein is correct.   (2)
4   Describe the role of mRNA in the process of protein synthesis.   (3)

Answers on page 26

Key Area 1.4
Proteins

Key points ⓘ

1 The variety of protein shapes arises from the sequence of amino acids in their molecules. ☐

2 The shape of protein molecules affects their function. ☐

3 The functions of proteins include **structural**, **enzymes**, **hormones**, **antibodies** and **receptors**. ☐

4 Structural proteins make up cell structures such as membranes. ☐

5 Enzymes function as biological **catalysts** and are made by all living cells. ☐

6 Hormones act as chemical messengers between cells and travel in body fluids such as blood. ☐

7 Antibodies are protein molecules that are involved in body defences. ☐

8 Receptors are found in cell membranes and recognise specific substances. ☐

9 Enzymes speed up cellular reactions and are unchanged in the process. ☐

10 The substance upon which an enzyme acts is called its **substrate**. ☐

11 The substance formed when an enzyme acts on its substrate is called the **product**. ☐

12 Some enzymes are involved in **degradation** reactions in which complex substrates are broken down into simple products. ☐

13 Some enzymes are involved in **synthesis** reactions in which complex products are built up from simple substrates. ☐

14 Enzyme molecules have **active sites** where they bind to substrate molecules and form an **enzyme-substrate complex**. ☐

15 Enzymes are **specific** because the shape of the active site of each enzyme molecule is complementary to the shape of its substrate. ☐

16 Each type of enzyme is most active in its **optimum** conditions. ☐

17 Enzymes and other proteins are affected by temperature and pH. ☐

18 Extremes of temperature or pH can result in changes to the molecular shapes of enzymes known as **denaturation**. ☐

19 When an enzyme is denatured, the shape of its active site no longer matches the shape of its substrate. ☐

20 A change in the molecular shape of an enzyme will affect its rate of reaction. ☐

21 A change in the molecular shape of any protein molecule will affect its function. ☐

Summary notes
Variety of proteins

There are many different proteins with different functions that are found in living organisms. The variety of shapes and functions arises from the different sequences of amino acids each protein has.

Functions of proteins

The functions of proteins can be grouped as shown in the table below.

Protein group	Function	Example
Structural units	Give strength and support to cellular structures	Membrane proteins give membranes their shape and strength Other structural proteins are found in muscle and bone
Hormones	Carry specific messages in the bloodstream of living organisms	Insulin signals to the mammal liver to store excess glucose
Antibodies	Provide specific defence against body invaders such as certain bacteria and viruses	Specific antibodies give defence against invading influenza viruses
Receptors	Allow cells to recognise specific substances	Liver cells have receptors for insulin
Enzymes	Act as biological catalysts to speed up chemical reactions in cells	Amylase speeds up the breakdown of starch molecules in animal digestive systems

> **Hints & tips** ★
>
> *Remember SHARE: Structural, Hormones, Antibodies, Receptors and Enzymes.*

> **Hints & tips** ★
>
> *There is more about hormones and receptors in Key Area 2.2b on page 48 and antibodies in Key Area 2.6 on page 65.*

Enzymes

Enzymes act as biological catalysts to speed up chemical reactions in cells, which would otherwise proceed too slowly to maintain life. The general action of enzymes is shown in Figure 1.12.

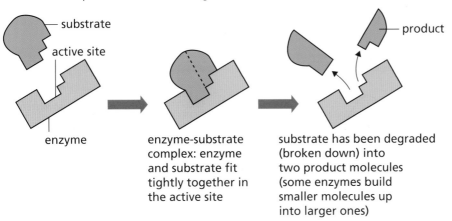

Figure 1.12 Enzyme action

The substance that a particular enzyme works on is called its substrate and those substances made are products. Enzyme molecules have a region of their surface known as the active site. The active site is complementary in shape to the substrate molecules and binds to them specifically to form a temporary enzyme-substrate complex; each enzyme has its own specific substrate.

Some enzymes are involved in degradation reactions in which complex substrates are broken down into simple products. Other enzymes are involved in synthesis reactions in which complex products are built up from simple substrates.

Different enzymes are responsible for different reactions because the substrates they work on differ in shape. Enzyme molecules are unchanged by their activity, so each enzyme molecule can catalyse a chemical reaction and then be used again.

Enzymes and temperature

Enzyme activity is affected by the temperature of its environment. Warmer conditions cause enzyme and substrate molecules to move around faster and so meet more regularly, resulting in increased rates of reaction. However, above a certain temperature enzyme molecules might become denatured. This process changes the shape of the active site and stops the denatured enzyme from working altogether. As shown in Figure 1.13, an enzyme has an optimum temperature at which it is most active.

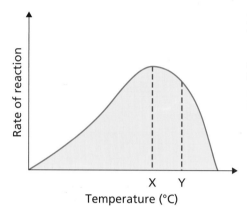

X is the optimum temperature for this reaction

At Y the higher temperature has started to denature the enzyme molecules, reducing the rate of reaction

Figure 1.13 The effect of temperature on enzyme action

Enzymes and pH

Enzyme activity is also affected by pH. Each enzyme has an optimum pH at which it is most active. Slight changes in the pH can result in changes to the active site, which will stop the enzyme working altogether (Figure 1.14).

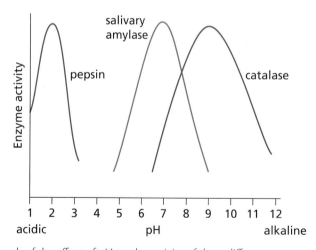

Figure 1.14 Graph of the effect of pH on the activity of three different enzymes

Measuring enzyme activity is a technique which you need to know for your exam.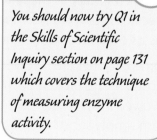

The table below shows the specific substrates and products of the degradation enzymes shown in Figure 1.14.

Enzyme	Substrate	Product(s)
Pepsin	Protein	Peptides and amino acids
Salivary amylase	Starch	Maltose
Catalase	Hydrogen peroxide	Oxygen and water

Hints & tips

Many students mix up the terms **specific** and **optimum**. Make sure you know the differences.

Hints & tips

Notice that enzyme activity can be measured by the rate at which products are made but could also be measured by the rate of disappearance of the substrate.

Hints & tips

You should now try Q1 in the Skills of Scientific Inquiry section on page 131 which covers the technique of measuring enzyme activity.

Key words

Active site – position on the surface of an enzyme molecule to which specific substrate molecules can bind

Antibody – protein that is involved in defence in animals

Catalyst – a substance that speeds up a chemical reaction by reducing the energy required to start it

Degradation – enzyme-controlled reaction in which a complex substrate is broken down into simple products

Denaturation – change in the shape of molecules of protein such as enzymes, resulting in them becoming non-functional

Enzyme – protein produced by living cells that acts as a biological catalyst

Enzyme-substrate complex – forms when an enzyme molecule binds to its substrate molecule before the product is made

Hormone – protein released by an endocrine gland into the blood to act as a chemical messenger

Optimum – conditions such as temperature and pH at which an enzyme is most active

Product – substance made by an enzyme-catalysed reaction

Receptor – cell surface protein which allows a cell to recognise specific substances

Specific – each different enzyme acts on one substrate only

Structural – referring to the proteins in membranes, muscle, bone, hair, nails etc.

Substrate – the substance on which an enzyme works

Synthesis – enzyme-controlled reaction in which simple substrates are built up into complex products

Questions

A Short-answer questions

1 State the feature of a protein that determines its molecular shape and the function it can carry out.   (1)
2 Give **two** different functions of proteins in living organisms.   (2)
3 State why enzymes are described as biological catalysts.   (1)
4 Give the effect of catalysing a reaction on an enzyme molecule's structure.   (1)
5 State why living cells require enzymes.   (1)
6 Give the term used for the substance on which an enzyme works.   (1)
7 Name the part of an enzyme molecule that binds to its substrate.   (1)
8 Give **two** different conditions that could lead to the denaturing of enzyme molecules.   (2)

B Longer-answer questions

1 Give the meaning of the term 'specific' as it is applied to enzymes.   (2)
2 Describe what happens to an enzyme molecule's structure and function when it becomes denatured.   (2)
3 Describe the main characteristics of enzymes.   (4)
4 The enzyme catalase breaks down hydrogen peroxide into water and oxygen as shown in Figure 1.15.

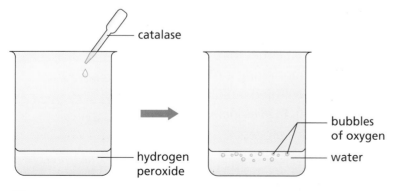

Figure 1.15
Suggest measurements which could be taken to show the rate of activity of catalase in this experiment.   (2)

Key Area 1.5
Genetic engineering

Key points

1 Normal control of cell activity depends on genetic information. ☐

2 Genetic information can be transferred artificially from the cells of one species to the cells of another by **genetic engineering**. ☐

3 Genetic engineering often, but not always, involves bacteria. ☐

4 Sections of chromosome can be transferred from one species to another, allowing the recipient to make new proteins. ☐

5 Genetic engineering is carried out by humans to allow a species to make a protein that is normally made by another species. ☐

6 Stages in genetic engineering include identifying and removing a section of DNA that contains the required gene from a **source chromosome**, inserting the required gene into an extracted bacterial plasmid, then inserting the **modified plasmid** into the **host bacterial cell**. ☐

7 Following genetic engineering, modified cells are cultured to produce a **genetically modified (GM)** strain or organism. ☐

8 The GM strain or organism can produce a new protein normally made by another species. ☐

9 Enzymes are used during genetic engineering to remove the required gene from the source chromosome, to cut open the bacterial plasmid and to seal the required gene into the bacterial plasmid. ☐

Summary notes
Genetic engineering

In recent decades, scientists have discovered that genetic material can be transferred artificially from cells of one species to the cells of another completely different species using techniques called genetic engineering. Human genes have been transferred to bacteria so that the fast-growing and rapidly reproducing bacteria make human proteins that are required medically. One example involves locating the gene that codes for the protein insulin on a human chromosome and then transferring it to a bacterial species. The modified bacteria can then be cultured rapidly to produce the large amounts of insulin required to treat an ever-increasing number of diabetics. Figure 1.16 shows the steps in this process.

Hints & tips ⭐

There is more about diabetes in Key Area 2.2b on page 48.

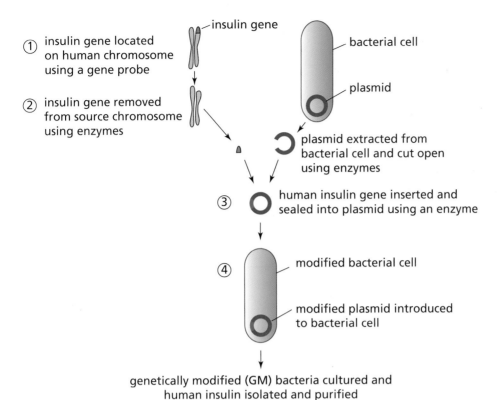

① insulin gene located on human chromosome using a gene probe

② insulin gene removed from source chromosome using enzymes

insulin gene

bacterial cell

plasmid

plasmid extracted from bacterial cell and cut open using enzymes

③ human insulin gene inserted and sealed into plasmid using an enzyme

④ modified bacterial cell

modified plasmid introduced to bacterial cell

genetically modified (GM) bacteria cultured and human insulin isolated and purified

Figure 1.16 Steps in genetic engineering

Genetically modified (GM) organisms

In recent times the transfer of genetic material from one organism to another has been applied more widely. Genes have been taken from a wide variety of organisms and used to transform cells of crop plants to improve yield. These are known as genetically modified (GM) crops. The table below shows some of the modifications that have been developed in crop plants in recent years.

Crop plant	Genetic modification and advantage
Rice	Genes from daffodil and soil bacteria allow the rice plant to produce vitamin A in its seeds
Rapeseed	Genes from other organisms that give the crop herbicide resistance allow herbicides to be used to kill weeds among the crop, so increasing yield
Tomato	Soil bacteria carry genes from other bacteria that code for an insecticide protein into tomato cells, so increasing yield
Potato	Genes from other organisms give potato plants fungal resistance, so increasing yield

Research is continuing into the use of artificial genes that can be engineered into domestic chickens to give them protection against bird flu.

GM organisms offer the potential for great benefits to humans but there are also issues connected to their production and use.

Hints & tips

There is more about GM crops in Key Area 3.5 on page 107.

Key words

Genetic engineering – the artificial transfer of genetic information from one donor cell or organism to another

Genetically modified (GM) – term used to describe a cell or organism that has had its genetic information altered, usually by adding a gene from another organism

Host bacterial cell – a bacterial cell that receives genetic material from a source chromosome

Modified plasmid – bacterial plasmid into which a required gene for a desirable characteristic has been inserted

Source chromosome – the chromosome from which the genetic material is obtained for transfer to another species

Questions ?

A Short-answer questions

1 Give the term applied to the transfer of DNA artificially between cells. (1)
2 Give the meaning of the term 'modified plasmid'. (1)
3 Give the meaning of the term 'GM organism'. (1)
4 Name **one** human hormone produced by genetic engineering. (1)
5 Arrange the following steps in a genetic engineering process into the correct order.
  A insert modified plasmid into host cell
  B extract the required gene
  C insert required gene into bacterial plasmid
  D identify DNA of required gene on donor chromosome (1)

B Longer-answer questions

1 Explain why bacteria are suitable organisms to use in genetic engineering. (2)
2 Genetic engineering often uses bacteria to produce the human hormone insulin. Outline the stages involved in this process. (4)

Answers on page 27

Key Area 1.6
Respiration

Key points

1   The chemical energy stored in **glucose** is released by all cells through a series of enzyme-controlled reactions called **respiration**. ☐

2   The energy released from the breakdown of glucose is used to generate **ATP**. ☐

3   The energy transferred by ATP can be used for cellular activities such as muscle cell contraction, cell division, protein synthesis and transmission of nerve impulses. ☐

4   In respiration, glucose is broken down to produce two molecules of **pyruvate** in the cytoplasm of cells. ☐

5   The breakdown of glucose to pyruvate releases enough energy to yield two molecules of ATP. ☐

6   If oxygen is available, **aerobic** respiration takes place in the mitochondria and each pyruvate is broken down to carbon dioxide and water, releasing enough energy to yield a large number of ATP molecules. ☐

7   When oxygen is not available, pyruvate remains in the cytoplasm and the **fermentation** pathway takes place. ☐

8   In the absence of oxygen, pyruvate is converted to **lactate** in animal cells and to **ethanol** and carbon dioxide in plant cells and in **yeast** cells. ☐

9   The breakdown of each glucose molecule via the fermentation pathway yields only the initial two molecules of ATP. ☐

10 Respiration begins in the cytoplasm. ☐

11 The process of fermentation is completed in the cytoplasm. ☐

12 Aerobic respiration is completed in the mitochondria. ☐

13 The higher the energy requirement of a cell, the greater the number of mitochondria present. ☐

14 The rate of respiration in living tissue is measured using a device called a **respirometer**. ☐

Summary notes
What is respiration?

The chemical energy stored in glucose is released by all cells through a series of enzyme-controlled reactions called respiration. The presence of energy in a foodstuff can be shown using the procedure illustrated in Figure 1.17. The energy is released by burning and the heat generated is used to raise the temperature of water.

The energy released from the breakdown of glucose is used to generate ATP. When the ATP is later broken down, its energy can be released and used in cellular

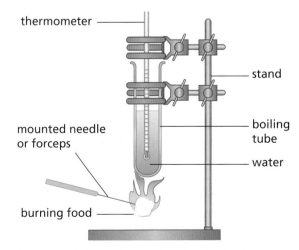

Figure 1.17 Release of energy from burning food

activities such as muscle contraction, cell division, protein synthesis and the transmission of nerve impulses. The role of ATP in the transfer of energy for muscle contraction is shown in Figure 1.18.

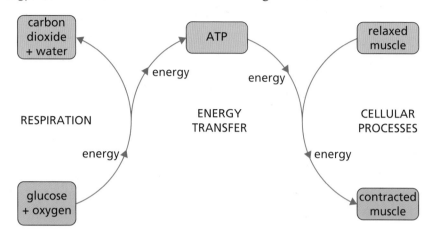

Figure 1.18 ATP energy transfer diagram

Cells that release a lot of energy to carry out their functions have relatively large numbers of mitochondria, which are the sites of aerobic respiration. A sperm cell, which has a high energy requirement because of its need to move rapidly, has many mitochondria in its middle region (Figure 1.19). There is more about sperm cells in Key Area 2.3.

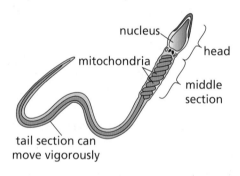

Figure 1.19 Many mitochondria in a sperm cell

> ### Hints & tips
> In your exam, you must be able to explain why certain cells require greater numbers of mitochondria.

Breakdown of glucose by aerobic respiration

In aerobic respiration, glucose molecules are broken down in two main stages.

1 Glucose is broken down to produce two molecules of pyruvate in the cytoplasm of cells. The breakdown of glucose to pyruvate releases enough energy to yield two molecules of ATP.
2 If oxygen is available, the pyruvate molecules enter the mitochondria and oxygen is used to complete the breakdown of glucose. Each pyruvate is broken down to the final products, carbon dioxide and water, releasing enough energy to yield a large number of ATP molecules.

> ### Hints & tips
> Remember that glucose is broken down to pyruvate in both fermentation and in the first stage of aerobic respiration.

This is summarised in Figure 1.20 and can be shown as:

glucose + oxygen → energy + carbon dioxide + water

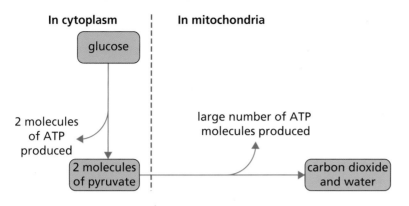

In cytoplasm | **In mitochondria**

glucose

2 molecules of ATP produced

large number of ATP molecules produced

2 molecules of pyruvate

carbon dioxide and water

Figure 1.20 Summary of aerobic respiration

Breakdown of glucose by fermentation

In the absence of oxygen, the fermentation pathway takes place. Fermentation occurs in the cytoplasm and does not require the presence of oxygen. In fermentation, the first step involves the breakdown of each glucose molecule to two molecules of pyruvate. In animal cells the pyruvate molecules are converted to lactate with the production of only the initial two molecules of ATP per glucose molecule respired. In plant and yeast cells the pyruvate molecules are converted to ethanol and carbon dioxide but again, only the initial two molecules of ATP are produced per glucose molecule respired. Fermentation in different cells is summarised in Figure 1.21. For plant and yeast cells this can be shown as **glucose → carbon dioxide + ethanol + energy** and for animal cells as **glucose → lactate + energy**.

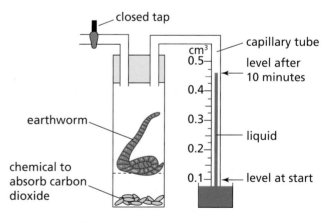

In animal cells | **In plant and yeast cells**

glucose — 2 ATP — 2 molecules of pyruvate — lactate

glucose — 2 ATP — 2 molecules of pyruvate — ethanol and carbon dioxide

Figure 1.21 Summary of fermentation in different cells

Respirometers

A respirometer is a device used to measure rates of respiration in living tissue. There are many types, but they usually measure oxygen uptake by tissue and are run for set periods of time to give respiration rates as consumption of oxygen per unit of time. A simple respirometer is shown in Figure 1.22. In this apparatus, oxygen used by the worm is measured as the coloured liquid moves up the scale to compensate. Any carbon dioxide produced by the worm is absorbed into the chemical and so does not affect the level of the coloured liquid.

closed tap

capillary tube

cm³

0.5 — level after 10 minutes

0.4

earthworm

0.3 — liquid

0.2

chemical to absorb carbon dioxide

0.1 — level at start

Figure 1.22 A simple respirometer

Using a **respirometer** is a technique which you need to know for your exam. **T**

Key words

Aerobic – in the presence of oxygen
ATP – adenosine triphosphate; a substance that transfers chemical energy in cells
Ethanol – alcohol produced as a result of fermentation of sugars by yeast
Fermentation – a type of respiration without oxygen
Glucose – simple sugar used as a respiratory substrate for the production of ATP
Lactate – substance produced during fermentation in animals and responsible for muscle fatigue
Pyruvate – substance produced by the breakdown of glucose in the cytoplasm in respiration
Respiration – a series of enzyme-controlled reactions resulting in the production of ATP from the chemical energy in glucose
Respirometer – a device to measure respiration rates in living tissue
Yeast – unicellular fungus used commercially in the brewing and baking industries

Questions ?

A Short-answer questions

1 Give the function of aerobic respiration in cells.   (1)
2 Name the substance which muscle cells break down to produce pyruvate.   (1)
3 Name the substance to which the energy released from the breakdown of glucose during aerobic respiration is transferred.   (1)
4 State the number of ATP molecules generated for each glucose molecule broken down in the absence of oxygen.   (1)
5 State the meaning of the term 'fermentation'.   (1)
6 Compare the number of ATP molecules generated during the breakdown of one glucose molecule by aerobic respiration and by fermentation.   (1)
7 State the location of fermentation in cells.   (1)
8 State the location of aerobic respiration in cells.   (2)
9 Name the end products of fermentation in the following types of cell:
   a) yeast   (1)
   b) animal   (1)
   c) plant.   (1)

B Longer-answer questions

1 Give the word equation for aerobic respiration.   (2)
2 Name the end products of aerobic respiration.   (3)
3 Describe the role of mitochondria in aerobic respiration.   (4)
4 Energy released during respiration is used for cellular activities.
   Give **three** examples of cellular activities that use this energy.   (3)
5 Compare aerobic respiration and fermentation in plant cells.   (4)
6 Explain why a sperm cell contains more mitochondria than a skin cell.   (2)
7 Describe how you would use a respirometer (Figure 1.22 on p. 23) to measure the rate of respiration in an earthworm. **T**   (2)

Answers on page 27

Answers

Key Area 1.1

A Short-answer questions

1 ribosome, cytoplasm, cell membrane
   [any 2]   (2)
2 nucleus, mitochondrion   [1 each]   (2)
3 plasmid   (1)
4 composed of different chemical substances/
   only plant cell walls are made of cellulose (1)
5 site of release of energy/ATP production in
   aerobic respiration   (1)
6 site of photosynthesis   (1)
7 site of protein synthesis   (1)
8 a) fluid-filled membrane-bound sac in a
       plant cell   (1)
   b) circular piece of DNA in a bacterial cell (1)

B Longer-answer questions

1 **similarities:** both have cell membranes; both
   have cytoplasm; both have ribosomes; both
   contain DNA
   differences: bacteria have cell walls but
   animal cells do not; bacterial cells have
   plasmids but animal cells do not; animal
   cells contain nuclei but bacterial cells do
   not; animal cells contain mitochondria but
   bacterial cells do not
   [any 5, 1 mark each]   (5)

2 made of cellulose; give cells support/prevent
   cell bursting   [1 each]   (2)
3 **similarities:** both have cell membranes; both
   have cell walls; both have cytoplasm; both
   have nuclei; both have mitochondria; both
   have ribosomes; both contain DNA
   [any 4, 1 mark each]
   differences: green plant cells have
   chloroplasts but fungal cells do not; green
   plant cell walls made of cellulose but fungal
   cell walls made of different substances/have
   different structures
   [any 1]   (5)
4 **similarity:** both have cell walls; both have
   ribosomes; both have cell membranes
   [any 1 similarity]
   difference: fungal cell has a nucleus, bacterial
   cell does not; cell walls are made of different
   substances; bacterial cells are smaller
   [any 1 difference]   (2)

Answers

Key Area 1.2

A Short-answer questions

1 phospholipid, protein   [1 each]   (2)
2 selectively permeable   (1)
3 the difference in concentration between
   two areas   (1)
4 diffusion **OR** osmosis   (1)
5 passive movement of molecules = 1
   from an area of high concentration to an area
   of lower concentration = 1   (2)
6 osmosis   (1)
7 osmosis refers to movement of water
   molecules only = 1
   in osmosis, the molecules move through a
   selectively permeable membrane = 1   (2)
8 movement of molecules from an area of
   low concentration to an area of higher
   concentration/against the concentration
   gradient; requires energy   [1 each]   (2)
9 plant roots taking up minerals from soil;
   nerve cells transporting sodium/potassium
   ions (other answers possible)   (1)
                                              ⇨

10 active transport requires energy but passive does not

OR active is against the concentration gradient but passive is down the gradient (1)

11 a) describes a plant cell with an enlarged vacuole filled with water (1)

b) describes a plant cell with a shrunken vacuole with its cell membrane pulled away from its cell wall (1)

12 water moves from an area of high concentration outside the cell to an area of lower water concentration inside the cell (1)

13 water moves from an area of high concentration inside the cell to an area of lower water concentration outside the cell (1)

14 vacuole shrunken, cytoplasm/cell membrane pulled from wall [1 each] (2)

B Longer-answer questions

1 allows them to lose waste materials/CO_2 from their cells; allows them to gain oxygen into their cells; allows them to gain nutrients/ glucose/amino acids into their cells [1 each] (3)

2 in osmosis, water molecules move through pores in protein molecules = 1 in active transport, protein molecules move molecules through the membrane = 1 (2)

3 potato cells would gain water by osmosis; from high water concentration in the beaker of pure water; to lower water concentration in the potato tissue; and the cells/tissue would become turgid/swell up [any 3] (3)

4 cell would lose water by osmosis; from high water concentration inside the cell; to lower water concentration in the sea water; and the cell would shrink (and die) [any 3] (3)

Answers

Key Area 1.3

A Short-answer questions

1 carries genetic information (1)
2 codes that allow proteins to be produced (1)
3 double-stranded helix (1)
4 A, G, T, C (1)
5 A pairs with T, G pairs with C [1 each] (2)
6 gene (1)
7 messenger RNA/mRNA (1)
8 amino acids (1)
9 ribosome (1)

B Longer-answer questions

1 double-stranded helix; two strands held together by (pairs of) bases; base pairs are complementary **OR** A pairs with T and G pairs with C [1 each] (3)

2 adenine/A specifically pairs with thymine/T; guanine/G specifically pairs with cytosine/C [1 each] (2)

3 mRNA has a complementary copy of the DNA base sequence; the sequence of bases on mRNA determines amino acid sequence in the protein produced [1 each] (2)

4 mRNA made in nucleus; carries complementary copy of DNA molecule; passes out of nucleus; moves to ribosome; amino acids assemble against the mRNA sequence of bases [any 3] (3)

Answers

Key Area 1.4

A Short-answer questions

1 sequence of amino acids (1)
2 structural units; enzymes; hormones; receptors; antibodies [any 2] (2)
3 speed up chemical reactions in living organisms (1)
4 molecular structure remains unchanged (1)
5 otherwise reactions would be too slow to maintain life (1)
6 substrate (1)
7 active site (1)
8 extremes of temperature or differences in pH [1 each] (2)

B Longer-answer questions

1 only works on one type of substrate; active site shape matches/is complementary to the shape of its substrate [1 each] (2)
2 active site shape becomes altered; can no longer fit its substrate [1 each] (2)
3 speed up reactions in cells; unchanged by their activity; are most active in optimum conditions; may be denatured by extremes of temperature or pH; specific to one substrate; made of protein [any 4] (4)
4 count number of bubbles/measure volume of oxygen produced; in a measured period of time/minute [1 each] (2)

Answers

Key Area 1.5

A Short-answer questions

1 genetic engineering/genetic modification (GM) (1)
2 plasmid that has been modified with DNA/required gene from another source (1)
3 a genetically modified organism with genes transferred from another species (1)
4 insulin; growth hormone [either] (1)
5 D, B, C, A (1)

B Longer-answer questions

1 have plasmids that can be removed and have donor genes added; accept modified plasmids; reproduce very rapidly [any 2] (2)
2 insulin gene located/identified; extracted from human chromosome; plasmid removed from bacterium; human gene inserted/sealed into plasmid; modified plasmid inserted into bacterium; bacterium cultured and insulin extracted; use of enzymes [any 4] (4)

Answers

Key Area 1.6

A Short-answer questions

1 to release energy from food (1)
2 glucose (1)
3 ATP (1)
4 2 ATP (1)
5 a type of respiration in the absence of oxygen (1)
6 more ATP made in aerobic respiration (1)
7 cytoplasm (1)
8 starts in cytoplasm; completed in the mitochondria [1 each] (2)
9 a) 2 ATP + ethanol + CO_2 (1)
   b) 2 ATP + lactate (1)
   c) 2 ATP + ethanol + CO_2 (1)
   $\Rightarrow$

B Longer-answer questions

1 glucose + oxygen → carbon dioxide + water + energy/ATP
   [all = 2; 1 missing part = 1]   (2)
2 CO_2, H_2O and ATP   [1 each]   (3)
3 site of aerobic respiration; breakdown of pyruvate; to CO_2 and H_2O; and release/production of ATP   [1 each]   (4)
4 muscle contraction; mitosis; protein synthesis; nerve transmission; active transport
   [any 3]   (3)
5 aerobic uses O_2, fermentation does not; aerobic produces a large number of ATP, fermentation produces two ATP; aerobic produces CO_2 and H_2O, fermentation produces CO_2 and ethanol; aerobic occurs in mitochondria, fermentation occurs in the cytoplasm   [1 each]   (4)
6 mitochondria are the sites of energy release (by aerobic respiration); sperm cells need more energy than skin cells because they have to swim/move (to eggs)   [1 each]   (2)
7 record the oxygen volume consumed/distance liquid moves up the scale; in a measured period of time/minute
   [1 each]   (2)

Practice assessment: Cell Biology (45 marks)

Section 1 (10 marks)

1 Which line in the table below correctly shows the structures present in a bacterial cell?

	Cell wall	Nucleus	Ribosome
A	✓	✓	✓
B	✓		✓
C		✓	✓
D	✓	✓	

2 Cell membranes are composed of
   A cellulose and starch
   C phospholipid and protein
   B cellulose and protein
   D starch and phospholipid.

3 The following statements refer to transport across membranes:
   1 against the concentration gradient
   2 down the concentration gradient
   3 requires ATP
   4 is passive
   Which statements apply to diffusion?
   A 1 and 2
   C 2 and 3
   B 1 and 3
   D 2 and 4

4 Which molecule carries a copy of the genetic code from the nucleus to where protein is assembled?
   A DNA
   C enzyme
   B ATP
   D mRNA

5 If a section of a DNA molecule contains 1000 bases of which 10% are adenine, how many of the bases would be cytosine?
   A 100
   C 400
   B 200
   D 800

6 Which term refers to the conditions under which an enzyme molecule is most active?
   A specific
   C denatured
   B optimum
   D complementary

7 The graph below shows changes in the enzyme and substrate concentrations in a seed over a period of time.

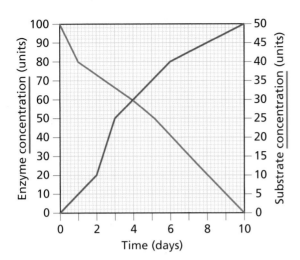

How many days does it take for the substrate concentration to decrease by 50%?

A 3   B 4

C 5   D 10

8 Which line in the table below correctly shows information about fermentation in muscle cells?

	Location	Products
A	Cytoplasm	2 ATP + lactate
B	Mitochondria	2 ATP + lactate
C	Cytoplasm	2 ATP + CO_2 and ethanol
D	Mitochondria	2 ATP + CO_2 and ethanol

9 The apparatus below was set up to investigate the effect of temperature on the respiration rate of earthworms over 10-minute periods.

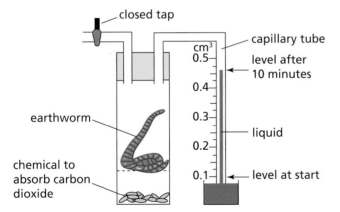

The experiment was repeated at three different temperatures and the rate of oxygen consumption by the worm calculated for each different temperature.

Which variable would have to be held constant for each repeat of the experiment?

A volume of oxygen consumed by the worm

B mass of worm used

C volume of coloured liquid in the capillary tube

D height of the wire platform in the glass tube

10 The diagram below shows the initial diameter of a disc of potato.

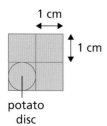

1 cm

1 cm

potato disc

The disc was placed in pure water for 1 hour.
Which of the diagrams shows the expected diameter of the disc after this time?

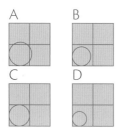

A   B

C   D

Section 2 (35 marks)

1 The diagram below shows a cell from human tissue. The cell has been magnified 500 times and the actual size of the magnified image is shown in the diagram. (1 mm = 1000 micrometres)

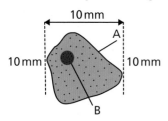

10 mm

A

10 mm   10 mm

B

   a) (i) Name structure A.   (1)
       (ii) Describe **one** way in which a fungal cell would differ from this cell.   (1)
   b) The cytoplasm of this cell contains few mitochondria.
      Explain what this suggests about the function of the cell.   (1)
   c) Calculate the actual size of the cell in micrometres.   (1SSI)

2 The diagram below shows a single plant cell immersed in a 10% sucrose solution. The concentration of the cell sap is equal to a 5% sucrose solution.

10% sucrose solution

5% sucrose solution

   a) Rewrite the sentence below, choosing the correct option from each set of brackets.
      Water would move (into/out of) this cell by osmosis and the cell would become
      (turgid/plasmolysed).   (1)

b) The graph below shows how the concentration of potassium ions within this cell changes as the oxygen concentration in its environment increases.

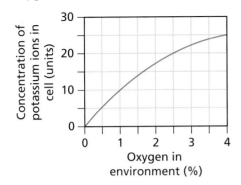

(i) Calculate the percentage increase in the potassium ion concentration in the cell when the oxygen concentration is 4% compared with 1%. (1SSI)

(ii) Name the process by which potassium ions enter the cell and explain the role of oxygen in this process. (2)

3 The diagram shows a small section of one strand of a DNA molecule, with letters representing bases.

DNA strand carrying bases of the genetic code

A G G C A G T C G C G G

a) Give the base sequence that would be expected on the DNA strand complementary to this part of the molecule. (1)

b) Explain how the base sequence on DNA determines the overall structure of the protein for which it codes. (1)

c) Name the part of a cell where proteins are assembled. (1)

4 The graph below shows how the rate of reaction of an enzyme was affected by an increase in temperature from 0 °C to 45 °C.

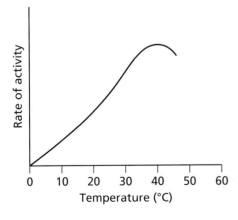

a) Describe how temperature has affected the rate of reaction of this enzyme over the range of temperature shown. (1SSI)

b) Predict how the rate of reaction would change as the temperature was raised to 60 °C, and give an explanation for your answer.

Prediction (1SSI)

Explanation (1)

c) Enzymes are proteins. Give **two** other functions of proteins. (2)

5 Genetic information can be transferred from one cell to another by genetic engineering.
   The diagram below shows stages by which a required gene from a human chromosome can be
   transferred artificially by genetic engineering to a bacterial cell.

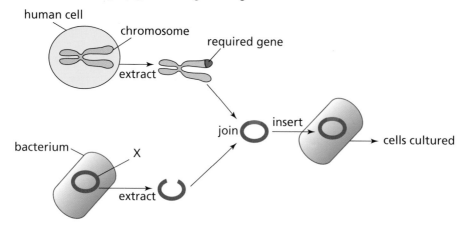

   a) Name structure X.   (1)
   b) Name **one** hormone produced by genetic engineering.   (1)
   c) Describe the role of enzymes in the process of genetic engineering as shown in the diagram
       above.   (3)

6 An investigation was carried out into the effect of temperature on the rate of respiration by yeast.
   Details of the apparatus, method used and results are given below.

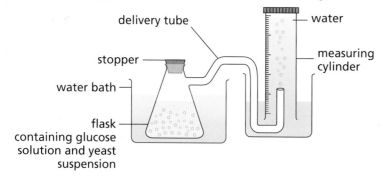

   1 Water baths were set up over a range of temperatures.
   2 $100\,cm^3$ of glucose solution and a 10% yeast suspension were allowed to reach the same temperature
      as the water bath.
   3 The glucose solution and the yeast suspension were mixed in the reaction flask as shown in the
      diagram.
   4 After 1 hour, the volume of gas in the measuring cylinder was measured.

Temperature (°C)	10	20	30	40	50
Volume of gas produced in one hour (cm³)	9	18	36	45	5

a) On a piece of graph paper, copy and complete the line graph to show the effect of temperature on the volume of gas produced in one hour over the range of temperatures. (2SSI)

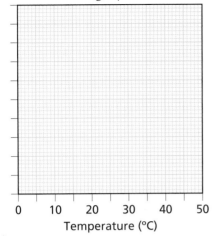

Temperature (°C)

b) Give **one** variable that would have to be kept constant at each temperature to ensure that a valid conclusion could be drawn. (1SSI)
c) Describe the relationship between the temperature and the rate of respiration. (1SSI)
d) Calculate, as the simplest whole number ratio, the volume of gas produced at 40 °C and the volume of gas produced at 10 °C. (1)
e) Describe the control flasks that would be set up to show that the gas was produced due to the activity of the yeast and to no other factor. (1SSI)

7 a) Describe the events which occur as a molecule of pyruvate undergoes aerobic respiration in a mitochondrion. (2)
b) Give **one** example of a cellular activity that requires the energy released during respiration. (1)

8 Read the following passage and answer the questions based on it.

DNA Profiles

Human DNA contains many base sequences common to all humans but some regions of a person's DNA are unique to that individual. These regions contain alleles that are inherited from the individual's mother and father. A person's DNA profile can be made by using a number of these unique regions.

In a paternity test a mother's DNA profile is compared with her child's to find which DNA she passed to them. The other half of the child's DNA can then be compared with an alleged father's DNA profile. If they don't match, this means he could not be the father of that child. If the DNA profiles match there is a probability of over 99% that he is the father. DNA profiles can be made from any human sample that contains cells with nuclei, such as saliva, semen, urine and hair. Profiles can be produced using samples that would be too small for other types of test. DNA resists degradation over long periods even after contamination with chemicals or bacteria. In forensics, DNA profiling is often used to eliminate suspects so that police are able to widen their investigations.

a) Describe how human DNA compares between different individuals. (1)
b) Describe what is meant by a paternity test. (1)
c) Explain why an alleged father cannot be identified with complete certainty using DNA profiles. (1)
d) Explain why samples of saliva or hair could contain DNA. (1)
e) Give a reason why older samples of saliva can still be useful for forensics. (1)

Answers to practice assessment: Cell Biology

Section 1

1 B, 2 C, 3 D, 4 D, 5 C, 6 B, 7 C, 8 A, 9 B, 10 A

Section 2

1 a) (i) cell membrane (1)
   (ii) fungal cell would have a cell wall **OR** vacuole (1)
   b) cell uses little energy/ATP (1)
   c) 20 micrometres (1)
2 a) water would move **out of** the cell and it would become **plasmolysed** (1)
   b) (i) 150% (1)
   (ii) active transport = 1
       oxygen allows aerobic respiration and release of energy/ATP = 1 (2)
3 a) TCCGTCAGCGCC (1)
   b) DNA base sequence determines the amino acids to be used **and** the order in which they occur (1)
   c) ribosome (1)
4 a) increasing temperature increases the rate of reaction up to 40 °C but further increases decrease the rate (1)
   b) prediction: reduced rate/rate decreases **OR** reaction stops (1)
      explanation: enzyme molecules denatured (1)
   c) structural; antibodies; hormones; receptors [any 2] (2)
5 a) plasmid (1)
   b) insulin/growth hormone (1)
   c) enzyme used to cut out required gene; enzyme used to cut open the bacterial plasmid/structure X; enzyme used to insert/seal gene into plasmid/structure X (3)

6 a) scales and labels; plotting and joining [1 each] (2)
   b) concentration of glucose solution/volume of yeast suspension/type, strain, species of yeast   [any 1] (1)
   c) as the temperature increases, the rate of respiration increases up to 40 °C then decreases (1)
   d) 5 : 1 (1)
   e) exactly the same apparatus, contents and conditions but without the yeast suspension/with dead yeast (1)
7 a) pyruvate acted upon by enzymes; converted to carbon dioxide and water; energy used to produce ATP   [any 2] (2)
   b) muscle contraction; mitosis; protein synthesis; nerve transmission; active transport   [any 1] (1)
8 a) many base sequences in common and some unique regions (1)
   b) checking that a child's DNA matches a father's (1)
   c) because the probability is less than 100% (1)
   d) they contain cells with nuclei (1)
   e) the DNA resists degradation over long periods (1)

Multicellular Organisms

Producing new cells

Key points !

1 The cell is the basic unit of life. ☐
2 The nucleus of a cell contains genetic information (DNA) organised into **chromosomes**. ☐
3 The **chromosome complement** of a cell is the number and type of chromosomes it contains. ☐
4 The chromosome complement of most cells is **diploid**, which means that their nuclei have two matching sets of chromosomes. ☐
5 **Mitosis** provides new cells for growth and repair of damaged cells. ☐
6 In mitosis the nucleus of a diploid parent cell divides to produce two diploid nuclei. ☐
7 Before mitosis starts, each chromosome undergoes DNA **replication** to form two **chromatids**. ☐
8 The sequence of events in mitosis starts with replicated chromosomes becoming visible as pairs of chromatids. A spindle of fibres forms, the chromosomes line up at the **equator** of the spindle and their chromatids are pulled to opposite **poles** by **spindle fibres** to form two new nuclei. ☐
9 After mitosis, the cell cytoplasm splits between the new nuclei to form two daughter cells. ☐
10 Daughter cells are genetically identical to their parent cell. ☐
11 Mitosis maintains the diploid chromosome complement to ensure that each daughter cell has all the genetic information it needs to carry out its functions. ☐
12 **Stem cells** are unspecialised cells in animals that can divide by mitosis to **self-renew**. ☐
13 Stem cells can be obtained from the embryo at a very early stage. In addition, tissue stem cells can be found in the body throughout life. ☐
14 Stem cells have the potential to become different types of **specialised** cell in multicellular animals. ☐
15 Stem cells are involved in the growth of animals. ☐
16 Stem cells are involved in the repair of damaged or diseased tissues in animals. ☐
17 Specialised cells in the bodies of multicellular organisms are organised into groups called **tissues**, **organs** and organ **systems**. ☐
18 There is a hierarchy within the bodies of multicellular organisms: cell → tissue → organ → system. ☐
19 A tissue is a group of similarly specialised cells carrying out the same broad function. ☐ ⇒

20 An organ is made up of a group of different tissues working together. ☐
21 Different organs carry out different functions. ☐
22 Organs are organised into groups called systems. ☐

Summary notes
Cells and chromosomes

The cell is the basic unit of life. Look back to Figure 1.1 on page 2 for a reminder of what individual cells look like.

A human being is an example of a multicellular organism with a body made up from cells. The nucleus of most cells contains two matching sets of chromosomes. A cell with two matching sets of chromosomes is said to be diploid. The diploid chromosome complement in humans is 46. Each chromosome has a group of inherited units called genes packaged into it. Genes are composed of a substance called deoxyribonucleic acid (DNA). DNA has the genetic information of the cell coded into its large, complex molecules. The table below summarises this information.

Term	Meaning
Nucleus	Organelle in a cell containing the diploid chromosome complement
Chromosome	Structure in which genetic information is packaged
Gene	Unit of genetic information found on a chromosome
Deoxyribonucleic acid (DNA)	Substance of which genes are composed and into which genetic information is coded

Mitosis and cell division

Living organisms grow by producing new cells. New cells are produced when the nuclei of parent cells divide by mitosis and their cytoplasm is split into two. Before mitosis, the DNA of a parent cell is copied exactly in a process called DNA replication. Following replication each chromosome appears as a double structure made up of two chromatids; each chromatid is a replicated chromosome. During mitosis, a fibrous structure called the spindle appears in the dividing cell. The chromosomes line up on the equator of the spindle and their chromatids are pulled apart by the spindle fibres. The chromatids are pulled towards the poles of the spindle where they form new nuclei.

Figure 2.1 shows a cell with a diploid chromosome complement of four undergoing mitosis and cytoplasm splitting. Each daughter cell is identical to the parent cell because it has an exact copy of the parent cell's genetic information.

Cell with diploid complement of four chromosomes

Chromosomes replicate and spindle fibres appear

Chromosomes move to the equator of the spindle

Chromatids pulled apart by spindle fibres

Nuclear membranes reform and cytoplasm starts to split to form two genetically identical daughter cells

Figure 2.1 Stages of mitosis in a cell with a diploid complement of four

Importance of mitosis

Mitosis ensures that each daughter cell contains a chromatid from each of the parent cell chromosomes. This means that each daughter cell is genetically identical to the parent cell and contains all the genetic information needed to carry out all of its activities and functions. Mitosis ensures that body cells of an individual organism are genetically identical to each other and that the diploid chromosome complement is maintained.

Specialisation of cells

In multicellular organisms, different cells become specialised to carry out different functions. Cells become specialised as their genetic material allows the production of specific proteins. These proteins are different in the various types of cell. Multicellular plants and animals have a variety of specialised cells (Figure 2.2).

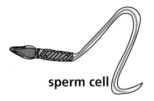

sperm cell

These cells carry half the genetic information. They have tails to swim towards the egg cell.

red blood cell

These cells are adapted to carry oxygen to other cells.

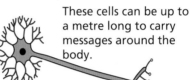

These cells can be up to a metre long to carry messages around the body.

neuron

muscle cell

These cells can change their length to help us move.

root hair cell

The root hair gives these cells a larger surface area to take in water from the soil.

palisade cell

These cells contain many chloroplasts to help the plant make food by photosynthesis.

Figure 2.2 Specialised cells in plants and animals

Specialisation of cells, in animals and plants, leads to the formation of a hierarchy of tissues, organs and systems.

Tissues

A living tissue is made from a group of cells with a similar structure and function, which all work together to do a particular job. Examples of animal tissues include muscle and nervous tissue. Plant tissues include xylem, phloem and epidermis.

Organs

An organ is made up of a group of different tissues working together to perform a particular function. Different organs carry out different functions. Examples of organs in animals include the brain, heart, lungs, stomach and skin. Plant organs are illustrated in Figure 2.3 and include the stem, root, flower and leaf.

Systems

A system is made up of a group of different organs that work together to do a particular job. Examples of organ systems in animals include the skeletal system, muscular system, digestive system, respiratory system, nervous system and circulatory system. Plant systems include the vascular system in the shoots and the roots. Figure 2.4 outlines the structure and function of systems in humans (a) and flowering plants (b).

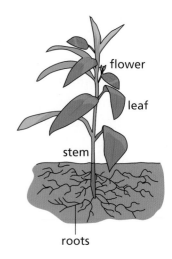

Figure 2.3 Plant organs

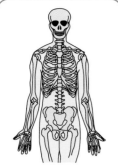

Skeletal system – provides structure to the body and protects internal organs

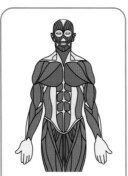

Muscular system – supports the body and allows it to move

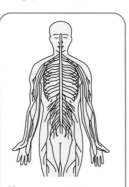

Nervous system – controls sensation, thought, movement, and virtually all other body activities

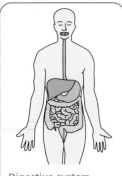

Digestive system – breaks down food and absorbs its nutrients

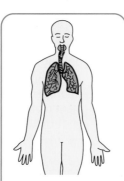

Respiratory system – takes in oxygen and releases carbon dioxide

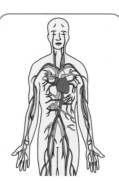

Circulatory system – transports oxygen, nutrients, and other substances to cells and carries away wastes

Figure 2.4(a) Systems in humans

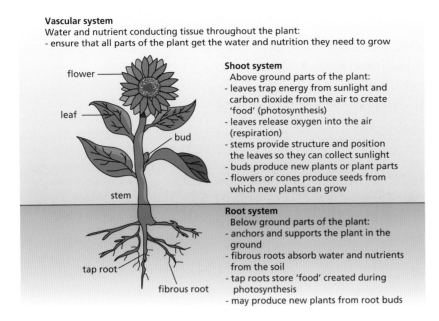

Vascular system
Water and nutrient conducting tissue throughout the plant:
- ensure that all parts of the plant get the water and nutrition they need to grow

flower
leaf
bud
stem
tap root
fibrous root

Shoot system
Above ground parts of the plant:
- leaves trap energy from sunlight and carbon dioxide from the air to create 'food' (photosynthesis)
- leaves release oxygen into the air (respiration)
- stems provide structure and position the leaves so they can collect sunlight
- buds produce new plants or plant parts
- flowers or cones produce seeds from which new plants can grow

Root system
Below ground parts of the plant:
- anchors and supports the plant in the ground
- fibrous roots absorb water and nutrients from the soil
- tap roots store 'food' created during photosynthesis
- may produce new plants from root buds

Figure 2.4(b) Systems in flowering plants

Hierarchy of organisation

Cells, tissues, organs and systems are often referred to as a hierarchy of organisation within the body of an animal or plant. The hierarchy of organisation in the human circulatory system and in the vascular system of a flowering plant are summarised in Figure 2.5.

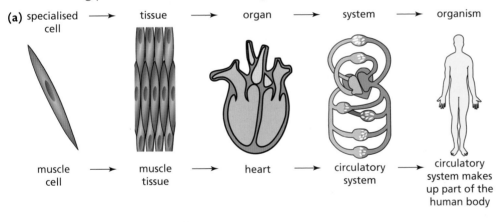

(a) specialised cell → tissue → organ → system → organism

muscle cell → muscle tissue → heart → circulatory system → circulatory system makes up part of the human body

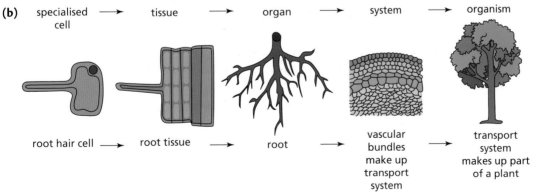

(b) specialised cell → tissue → organ → system → organism

root hair cell → root tissue → root → vascular bundles make up transport system → transport system makes up part of a plant

Figure 2.5 Hierarchy of organisation in **(a)** the human circulatory system and **(b)** the vascular system in a flowering plant

Stem cells

Stem cells in animals are unspecialised cells which can divide by mitosis to self-renew. This produces new cells for growth and repair of damaged tissue. They also have the potential to become different cell types, specialised to carry out particular functions. Stem cells can be obtained from the embryo at a very early stage. In addition, tissue stem cells can be found in the body throughout life.

Stem cells in the human body produce many different types of specialised cell, which include muscle cells, red blood cells and neurons as shown in Figure 2.6.

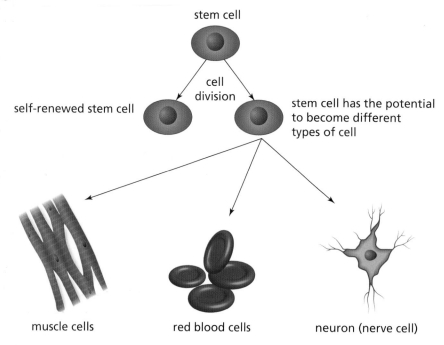

Figure 2.6 The self-renewal and potential for specialisation of stem cells

Key words

Chromatid – replicated copy of a chromosome visible during cell division

Chromosome – structure containing hereditary material; composed of DNA that codes for all the characteristics of an organism

Chromosome complement – the characteristic number of chromosomes in a typical cell of an organism

Diploid – describes a cell containing two sets of chromosomes

Equator – middle position of a spindle where chromosomes are attached during mitosis

Mitosis – division of the nucleus of a cell that leads to the production of two genetically identical diploid daughter nuclei

Organ – a group of different tissues that work together to carry out a particular function, e.g. heart and lungs

Poles – opposite ends of a spindle to which chromatids migrate during mitosis

Replication – copying of DNA to produce chromatids before mitosis

Self-renew – property of stem cells by which they divide by mitosis to produce more stem cells

Specialised – description of a cell that has become differentiated to carry out a particular function

Spindle fibres – threads produced during mitosis to pull chromatids apart

Stem cell – unspecialised cell capable of dividing into cells that can develop into different cell types

System – a group of organs that work together to carry out a particular function, e.g. circulatory, respiratory and digestive

Tissue – a group of similar cells carrying out the same function

Questions ?

A Short-answer questions

1 Give the term applied to a group of similar cells that carry out the same function.　(1)
2 Give **one** example of a tissue from each of the following organisms:
　a) plant　　　　　　　　　　　b) human.　(2)
3 Give the meaning of the term 'organ'.　(1)
4 Give **one** example of an organ from each of the following organisms:
　a) plant　　　　　　　　　　　b) human.　(2)
5 Give the meaning of the term 'system'.　(1)
6 Name **two** different human body systems.　(2)
7 Give the term used to describe a cell containing two matching sets of chromosomes.　(1)
8 Give the term applied to replicated chromosomes before they separate during mitosis.　(1)
9 Describe the role of the spindle fibres during mitosis.　(2)
10 Explain what is meant by the term 'specialised', as applied to cells.　(1)
11 Give **two** properties of stem cells.　(2)
12 Give **two** types of human cell which can be formed when a stem cell specialises.　(2)
13 Describe what happens to the chromosomes of a cell before mitosis takes place.　(2)
14 Explain the importance of maintaining the diploid chromosome complement following mitosis.　(1)
15 'A stem cell has the potential to become different types of cell.' Explain the meaning of this statement.　(2)

B Longer-answer questions

1 Describe the hierarchy of organisation in the human body.　(4)
2 Describe the events of mitosis in a diploid body cell.　(4)
3 Explain the importance of mitosis to multicellular organisms.　(3)
4 Describe the changes to a human stem cell that result in the production of a red blood cell.　(3)

Hints & tips

For help with question 4, read page 69 for details of the structure of a red blood cell.

Answers on page 75

Control and communication: nervous control

Key points ⚠

1 The parts of the nervous system are the **brain**, the **spinal cord** and the **nerves**. ☐

2 The nervous system is made up of cells called **neurons**. ☐

3 There are three types of neuron: **sensory, inter** and **motor**. ☐

4 The **central nervous system (CNS)** is made up of the brain and spinal cord. ☐

5 The brain is made up of the **cerebrum**, **cerebellum** and **medulla**. ☐

6 The cerebrum is responsible for conscious thoughts, reasoning, memory and emotions. ☐

7 The cerebellum controls balance and coordination. ☐

8 The medulla controls involuntary functions such as breathing, heart rate and peristalsis. ☐

9 The CNS processes information from sense organs and coordinates responses. ☐

10 Sense organs contain **receptors** which generate **electrical impulses** in response to sensory inputs called **stimuli**. ☐

11 Electrical impulses carry messages along sensory neurons to inter neurons within the CNS. ☐

12 Electrical impulses in inter neurons within the CNS carry messages to motor neurons. ☐

13 Electrical impulses in motor neurons carry messages to **effectors** such as muscles and glands, which bring about a response. ☐

14 Responses from muscles are usually rapid and those from **glands** are usually slower. ☐

15 A **synapse** is a narrow gap between the neurons. ☐

16 Chemicals are released into the synapse from the neuron leading to it. These chemicals allow an electrical impulse to be generated in the next neuron after the synapse. ☐

17 **Reflexes** are very rapid actions since they involve only a few neurons and do not require complex processing of information by the brain. ☐

18 Reflex actions help to protect the body from harm. ☐

19 The group of neurons involved with the structure and function of a reflex action is called a **reflex arc**. ☐

Summary notes
Control and communication

Internal communication and control is required for the survival of a multicellular organism. This is because cells in the whole organism cannot function independently of each other. In animals there are two broad types of control, nervous and hormonal (see Key Area 2.2b on page 48).

Nervous control in animals

The nervous system of mammals has two main parts: the central nervous system (CNS), which consists of the brain and spinal cord, and the nerves (neurons), which connect the CNS to all parts of the body (Figure 2.7).

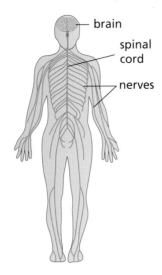

Figure 2.7 Parts of the human nervous system

The nervous system receives stimuli, which are pieces of sensory information about the external and internal environment of the body. Stimuli are detected by receptors. Input of a stimulus causes an electrical impulse which carries a message into the central nervous system where it is processed. The CNS sorts out messages received from receptors and sends electrical impulse messages along nerves to effectors. The effectors are the muscles or glands that make the appropriate response (Figure 2.8). A response to a stimulus can be a rapid action from a muscle or a slower response from a gland.

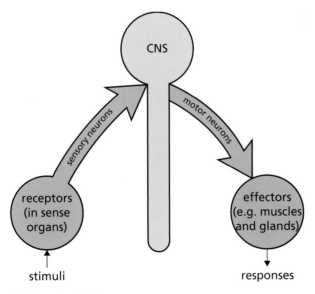

Figure 2.8 Flow of information in the nervous system

Cells of the nervous system

The nervous system is made up from fibre-shaped cells called neurons. Nerves are bundles of neurons. Neurons are the longest cells of the body. There are three main types (Figure 2.9).

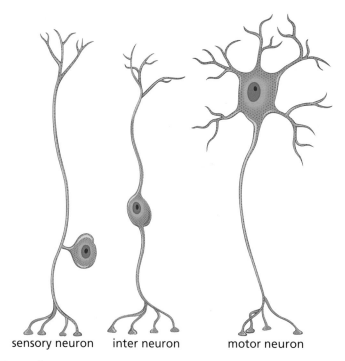

Figure 2.9 Types of neuron

Sensory neurons carry messages from the receptors in the sense organs towards the CNS. In the CNS there are shorter inter neurons that connect sensory neurons to motor neurons. Motor neurons carry electrical messages from the CNS to the muscles or glands that are the effectors. Motor neurons enable the effectors to make a response, which can be an action from a muscle or a gland.

Synapses

A synapse is a gap between neurons. Chemicals released from tiny vacuoles at the end of one neuron diffuse across the synapse to generate an electrical impulse in the next neuron (Figure 2.10).

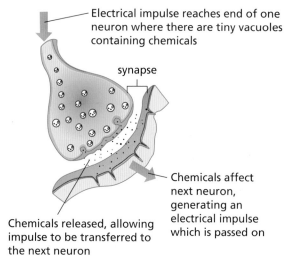

Figure 2.10 A synapse

Rapid reflex action

Reflexes are rapid, automatic and protective actions (see the table below). They are automatic responses to damaging stimuli. We do not have to learn or think about the response because that could cause delay and lead to further harm to the body.

Reflex action	Protective value
Cough	Clears the windpipe
Sneeze	Clears nasal passages
Blink	Clears the eye surfaces
Withdrawal	Removes body from extreme heat and pain

The reflex arc

The table below shows the pathway of a withdrawal reflex action from the detection of a heat stimulus to the withdrawal response.

Stage in withdrawal reflex	Description of stage
Stimulus	The touch of a hot object
Sense organ	Heat receptor cells in skin detect the stimulus
Sensory neuron	Messages carried as electrical impulses to the CNS
Inter neuron	Messages passed as electrical impulses through the CNS to motor neurons
Motor neuron	Messages carried as electrical impulses to the muscle
Muscle	Muscle contracts
Response	Withdrawal from the extreme heat

Figure 2.11 shows the reflex arc as a diagram.

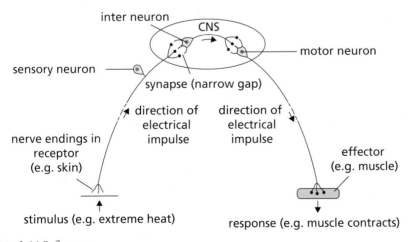

Figure 2.11 Reflex arc

The brain

The human brain is an extremely complex organ, made up of hundreds of billions of neurons. It can be divided into three different regions (Figure 2.12).

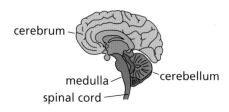

Figure 2.12 Side view of the brain sliced down the middle

Each region has specific functions associated with it, as outlined in the following table.

Region	Function
Cerebrum	Site of conscious thoughts, reasoning, memory and emotions
Cerebellum	Centre of control for balance and coordination of movement
Medulla	Centre of control for breathing and heart rate

Key words

Brain – organ of the central nervous system of mammals where vital functions are coordinated

Central nervous system (CNS) – part of the nervous system made up of the brain and spinal cord

Cerebellum – part of the brain that controls balance and coordination of movement

Cerebrum – large folded part of the brain that controls conscious thoughts, reasoning, memory and emotion

Effectors – muscles or glands which make the response in a reflex arc

Electrical impulses – method of carrying messages along neurons

Gland – group of cells which can synthesise and release a substance such as a hormone

Inter neuron – nerve cell that transmits electrical impulses from sensory neurons to motor neurons within the CNS

Medulla – part of the brain controlling breathing and heart rate

Motor neuron – nerve cell that carries electrical impulses from the CNS to effectors such as muscles or glands

Nerves – bundles of neurons that connect receptors to the CNS and the CNS to the effectors

Neuron – nerve cell that is specialised to transmit electrical impulses

Sensory neuron – nerve cell that transmits electrical impulses from a sense organ to the CNS

Receptor cell – cell that can detect stimuli inside or outside the body

Reflex – a rapid action which protects the body from harm

Reflex arc – pathway of information from a sensory neuron through an inter neuron directly to a motor neuron

Spinal cord – part of the central nervous system of a mammal that runs within its backbone

Stimuli – changes in the environment detected by receptor cells that trigger a response in an organism (*sing.* stimulus)

Synapse – gap between two neurons in which chemicals transfer messages

Questions ?

A Short-answer questions

1 Name the **two** parts of the human central nervous system (CNS). (2)
2 Describe the role of the CNS in humans. (2)
3 Describe how information is passed along a neuron. (1)
4 State how neurons are linked in the nervous system. (1)
5 State the general function of receptors as part of the nervous system. (1)
6 Give the importance of the rapid reflex action in humans. (1)

B Longer-answer questions

1 Describe the functions of each of the three main regions of the brain. (3)
2 Describe the role of each of the three types of neuron in the nervous system. (3)
3 Describe what happens when an electrical signal arrives at a synapse. (3)
4 Describe the flow of information in a reflex arc. (4)

Answers on page 76

Key Area 2.2b
Control and communication: hormonal control

Key points !

1 **Endocrine glands** release hormones into the bloodstream. ☐
2 Hormones are proteins that act as chemical messengers. ☐
3 Hormones affect **target tissues**, which contain cells with complementary **receptor proteins** to detect them. ☐
4 Receptors are complementary so that only target tissues are affected by specific hormones. ☐
5 The **pancreas** monitors the blood glucose concentration in the blood. ☐
6 The pancreas produces the hormones **insulin** and **glucagon**. ☐
7 Insulin is released in response to increased blood glucose concentration and stimulates the **liver** to convert excess glucose to **glycogen** for storage. ☐
8 Glucagon is released in response to decreased blood glucose concentration and stimulates the liver to release glucose from stored glycogen. ☐

Summary notes
Hormonal control in animals

Endocrine glands release hormones into the bloodstream. Hormones are chemical messengers which are transported in the blood to their target organ. Target organs have cells with complementary receptor protein molecules on their surfaces as shown in Figure 2.13. Although hormones are carried to all the body organs, they only affect their specific target tissues, which respond to them.

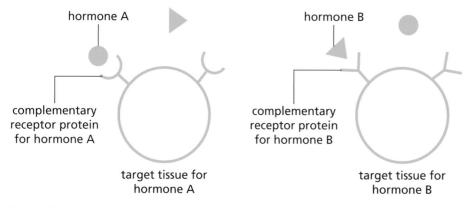

hormone A

complementary receptor protein for hormone A

target tissue for hormone A

hormone B

complementary receptor protein for hormone B

target tissue for hormone B

Figure 2.13 Hormones and their target tissues

Since hormones are transported in the blood, hormonal control is slower than nervous control.

Blood glucose regulation in humans

Eating carbohydrates increases the blood glucose concentration. A rise in blood glucose concentration is detected by receptor cells in the pancreas. This causes the pancreas to produce the hormone insulin. Insulin travels in the blood to its target tissues in the liver, and causes liver cells to take up glucose and store it as glycogen. This reduces the blood glucose concentration.

Missing a meal or taking a lot of exercise can result in a decrease in the blood glucose concentration. A decrease in blood glucose concentration is detected by different receptor cells in the pancreas. This causes the pancreas to produce the hormone glucagon. Glucagon travels in the blood to its target tissues in the liver, and causes the liver cells to convert glycogen to glucose. The glucose is released, increasing the blood glucose concentration. This is summarised in Figure 2.14.

> **Hints & tips** ★
>
> You can see the position of the pancreas and liver in Figure 2.33 on page 73.

> **Hints & tips** ★
>
> **GLU**ca**GON** is needed when **GLU**cose is **GON**e.

> **Hints & tips** ★
>
> A useful rhyme:
> Low blood sugar, glucose gone; what you need is glucagon
>
> To turn glucose into glycogen, what you need is insulin

> **Hints & tips** ★
>
> This is an occasion when spelling is very important. You must not confuse glycogen and glucagon.

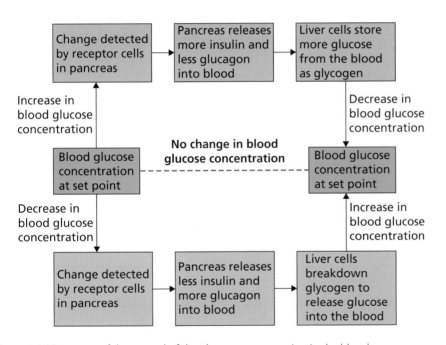

Figure 2.14 Summary of the control of the glucose concentration in the blood

In recent years there has been a huge increase in the number of people in Scotland with diabetes.

Diabetes

Diabetes is an example of a communication pathway that has failed. It is caused by failure to control glucose levels in the body. There are two forms of the disease:

● **Type 1 diabetes** is a severe insulin deficiency, which usually appears in childhood. Until the discovery of insulin in 1922, diabetes was a fatal disease. Today diabetes can be treated by injections with insulin and by careful diet. Insulin can be extracted from the pancreas tissue of cattle and pigs, or it can be made in fermenters by genetically modified bacteria. Treatment is improving all the time, with the development of fast-acting and slow-acting insulin preparations, simpler injection pens, oral insulin preparations, portable insulin infusion pumps and even the possibility of tissue transplants and stem cell therapy in the future. In Key Area 1.5, the part played by genetic engineering in the production of the artificial insulin needed to treat diabetes is explained.

- **Type 2 diabetes** occurs because, although insulin is made, insulin receptors in the target tissues do not respond and so the hormone has no effect. The treatment of this form of diabetes involves regulation of blood glucose through strict control of diet. Type 2 diabetes tends to appear in overweight people and is increasing. It accounts for 90% of diabetes cases in Scotland and is linked to lifestyle choices involving a diet leading to obesity.

Key words

Endocrine gland – gland that produces and releases a hormone directly into the blood
Glucagon – hormone produced by the pancreas, responsible for triggering the conversion of glycogen into glucose in the liver
Glycogen – animal storage carbohydrate located in the liver and muscle tissues
Insulin – hormone produced by the pancreas that triggers the conversion of glucose into glycogen in the liver
Liver – large organ, beside the stomach, with many important functions including a role in blood glucose control
Pancreas – organ responsible for the production of digestive enzymes and the hormones insulin and glucagon
Receptor protein – molecule which is complementary to a specific hormone molecule
Target tissue – tissue with receptor protein molecules on its cell surfaces that are complementary to a specific hormone

Questions ?

A Short-answer questions

1 State the function of endocrine glands. (1)
2 Give **one** example of an endocrine gland in humans. (1)
3 State how hormones carry their messages around the body. (1)
4 Give the term applied to tissues carrying receptor proteins complementary to a specific hormone. (1)
5 Name **two** hormones involved in the regulation of blood glucose concentration. (2)

B Longer-answer questions

1 Describe the specific action of hormones. (4)
2 Describe the control of blood glucose following an increase in blood glucose concentration in humans. (4)

Answers on page 76

Reproduction

Key points

1 Most cells in the bodies of most multicellular organisms are diploid. ☐
2 Multicellular organisms produce sex cells called **gametes**, which are **haploid**. ☐
3 Gametes are either male or female. ☐
4 In flowering plants, gametes are produced in the flowers. ☐
5 Male gametes in flowering plants are inside **pollen grains** made in the **anthers** of flowers. ☐
6 Female gametes in flowering plants are inside **ovules** made in the **ovaries** of flowers. ☐
7 In male animals, the gametes are called **sperm cells** and are made in **testes**. ☐
8 Sperm cells are very small and are said to be **motile** because they can swim independently using the whipping action of their tails. ☐
9 In female animals, the gametes are called **egg cells** and are made in **ovaries**. ☐
10 Egg cells are larger cells with a food supply. ☐
11 In both animals and plants, **fertilisation** involves the fusion of the haploid gamete nuclei to form a diploid **zygote**, which then divides to form an **embryo**. ☐

Summary notes
Chromosomes, gametes and reproduction

Most cells in the bodies of most multicellular organisms are diploid, which means that they contain two matching sets of chromosomes. Multicellular organisms produce sex cells called gametes, which are haploid. Haploid cells contain one set of chromosomes. Gametes are either male or female.

Reproduction in flowering plants

In flowering plants, gametes are produced in the reproductive organs in flowers. Male gametes are inside the pollen grains produced in the anthers of flowers. Female gametes are inside the ovules produced in the ovaries. These structures are shown in Figure 2.15.

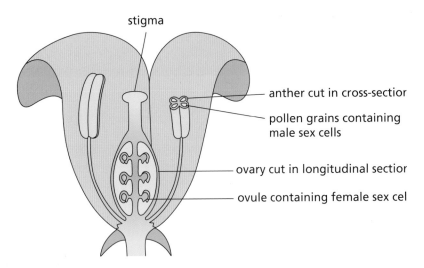

Figure 2.15 Flower structure

During the reproduction process, fertilisation occurs when the nucleus of a pollen grain joins with a nucleus in the ovule. The fusion of the haploid gametes produces a diploid zygote. After fertilisation, the female parts of the flower develop into a fruit. The fertilised ovule becomes a zygote which divides to form an embryo within a seed. The ovary wall becomes the rest of the fruit.

Reproduction in animals

In animals, gametes are produced in specialised organs. Figure 2.16 shows the reproductive organs in human males and females. Male gametes are sperm cells and are produced in the testes. Female gametes are called egg cells (ova) and are produced in the ovaries. Sperm cells have tails that allow them to swim to egg cells. The egg cells have a food supply. Figure 2.17 shows the structure of sperm and egg cells. Fertilisation occurs when a sperm cell nucleus joins with an egg cell nucleus to form a diploid zygote. After fertilisation, the food supply from the egg provides energy to allow the zygote to divide to form an embryo.

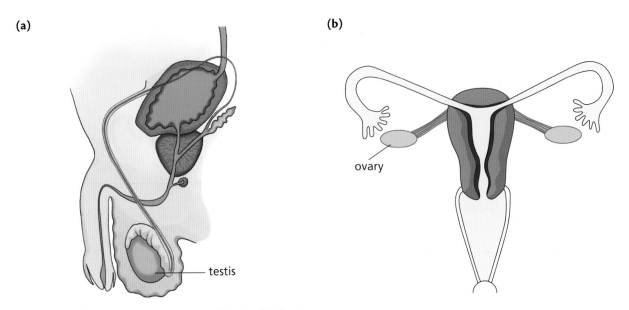

Figure 2.16 Human reproductive organs: (a) male, (b) female

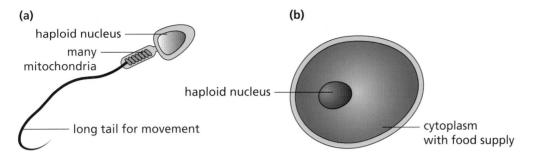

(a)
- haploid nucleus
- many mitochondria
- long tail for movement

(b)
- haploid nucleus
- cytoplasm with food supply

Figure 2.17 Human gametes: **(a)** male sperm cell, **(b)** female egg cell

In both plants and animals, the fusion of the nuclei of haploid gametes produces a diploid zygote. Figure 2.18 shows fertilisation and the formation of an embryo in humans.

Hints & tips ⭐

There is more about chromosome number and cell division in Key Area 2.1 on p. 35.

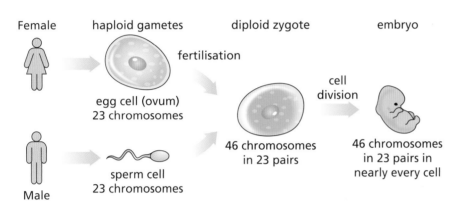

Female — haploid gametes — diploid zygote — embryo

fertilisation

egg cell (ovum)
23 chromosomes

sperm cell
23 chromosomes

Male

cell division

46 chromosomes in 23 pairs

46 chromosomes in 23 pairs in nearly every cell

Figure 2.18 Fertilisation and embryo formation in humans

Key words

Anther – organ within a flower that produces pollen grains
Egg cell – female gamete produced by ovaries in animals
Embryo – structure produced when a zygote divides repeatedly
Fertilisation – the fusion of the nuclei of haploid gametes
Gamete – sex cell containing the haploid chromosome complement
Haploid – describes a cell, usually a gamete, containing one set of chromosomes
Motile – able to move under its own power
Ovaries – female sex organs (*sing.* ovary)
Ovule – structure containing a female gamete, produced by ovaries in plants
Pollen grain – structure produced in the anthers of a flower that contains the male gamete
Sperm cell – gamete produced in the testes of male animals
Testes – male sex organs in animals for the production of sperm
Zygote – fertilised egg cell

Questions ?

A Short-answer questions

1 Give the meanings of the terms 'haploid' and 'diploid'. (2)
2 Name the sites of gamete production in plants. (2)
3 Name the male and female gametes in plants. (2)
4 Name the sites of gamete production in animals. (2)
5 Name the male and female gametes in animals. (2)
6 Give the term used for a fertilised egg cell. (1)
7 Describe **two** differences between the structure of a sperm cell and an egg cell. (2)

B Longer-answer questions

1 Explain why body cells are diploid and gametes are haploid. (2)
2 Describe the cells and processes involved in reproduction in flowering plants. (4)
3 Describe the cells and processes involved in reproduction in animals. (4)

Answers on page 76

Key Area 2.4
Variation and inheritance

Key points !

1 Differences that can be seen between individual members of a species are called **variation**. ☐

2 Combining genes from two parents contributes to the variation within a species. ☐

3 In **discrete** variation, the clear-cut differences in a characteristic involve **single gene inheritance** and show a discontinuous pattern in which measurements fall into distinct groups. ☐

4 In **continuous variation** there is a range of variation between the extremes of a characteristic. ☐

5 Discrete variation arises because individuals carry different **alleles** of the same gene. ☐

6 In continuous variation, the characteristic is usually **polygenic**, which means that different alleles of many different genes are involved. ☐

7 In polygenic inheritance, the characteristics show a range of values between a minimum and a maximum. ☐

8 The **phenotype** of an organism is the outward appearance of a characteristic it has. ☐

9 The **genotype** of an organism is a statement of the alleles it has for a particular characteristic. ☐

10 **Dominant** alleles always show in the phenotype of an organism that has them. ☐

11 **Recessive** alleles only show themselves in the phenotype if an organism has inherited one from each parent. ☐

12 A **homozygous** organism has two copies of the same allele of a particular gene; one from each parent. ☐

13 A **heterozygous** organism has inherited two different alleles of the same gene; one from each parent. ☐

14 A **monohybrid cross** can be set up to study the inheritance of a pair of contrasting characteristics, such as yellow and green seed colour in pea plants, which are caused by alleles of the same gene. ☐

15 A monohybrid cross starts with a **parental generation (P)** producing a **first generation (F_1)** and can be carried through to a **second generation (F_2)** by crossing members of the F_1. ☐

16 Predicted phenotype ratios among offspring are not always achieved because of the random nature of fertilisation. ☐

17 **Punnett square diagrams** can be used to explain inheritance patterns. ☐

18 **Family tree diagrams** can be drawn and genotypes can be identified from them. ☐

Summary notes
Discrete and continuous variation

Differences that can be seen between individual members of a species are called variation. Combining genes from two parents contributes to the variation within a species.

In discrete variation, the differences are clear-cut and show a discontinuous pattern. A good example is the human ability to roll the tongue. An individual can either do it or is unable to; it is clear-cut. Discrete variation arises through single gene inheritance in which individuals carry different discrete forms of the gene for the characteristic. A form of a gene is called an allele.

In continuous variation there is a range of variation between a minimum and a maximum. The characteristic is usually polygenic, which means that many different genes are involved. Most features of an individual's phenotype are polygenic and show continuous variation. A good example is human height, which shows a range from very short to very tall individuals with all possible heights in between.

Figure 2.19 shows examples of the patterns of discrete and continuous variation.

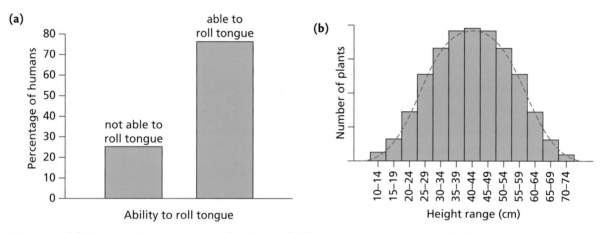

Figure 2.19 (a) Discrete variation – tongue-rolling bar graph **(b)** Continuous variation – height histogram

What is a gene?

A gene is a section of DNA carrying genetic information to make a protein. It carries the information to create a particular characteristic. Genes exist in different forms called alleles, which give different versions of the same characteristic. Alleles are given symbols, usually letters of the alphabet, to make working with them easier. The table below shows some examples of organisms and the alleles of one of their genes.

> **Hints & tips** ⭐
>
> *It is essential to appreciate the difference between a gene and an allele. An allele is a form of a gene – humans have **blood group genes** but group A is caused by an **allele**.*

Organism	Gene	Allele symbol	What the allele does in the organism
Human	Tongue-rolling	R	Allows the tongue to be rolled up from the sides
		r	Tongue cannot be rolled up from the sides
Fruit fly	Wing length	W	Normal length of wings
		w	Very short wings – they cannot fly
Pea plant	Seed colour	Y	Pea seeds yellow coloured
		y	Pea seeds green coloured

Genetics and its language

Genetics is the study of variation and inheritance and it has its own language, which is best understood through an example. In the early nineteenth century an Austrian monk called Gregor Mendel studied monohybrid inheritance in garden pea plants. He noted that the plants produced yellow or green peas and started experiments to study how pea seed colour was inherited. Mendel was able to show that each individual pea plant had two sets of genetic information: a set from each of its two parents. This meant that it had two alleles of each gene; these could be the same or different. A plant with two alleles the same is said to be homozygous while a plant with different alleles is heterozygous. The alleles the plant has form its genotype but the appearance of the plant, for example its seed colour, is its phenotype. An individual plant passes on only one of its two alleles into gametes to be inherited by offspring.

Mendel also showed that alleles could be dominant or recessive; dominant alleles always showed up in the phenotype of a plant while recessive alleles only showed up if a plant had two copies. It turned out that yellow seed colour was dominant to green seed colour. We give dominant alleles a capital letter symbol and recessive alleles a lower case letter symbol. Homozygous plants always breed true, which means that their characteristics always appear in their offspring's phenotype. Heterozygous individuals do not always breed true and their offspring might have different phenotypes. The table below summarises this information in relation to pea seed colour.

Genotype	Phenotype	Homozygous or heterozygous	Description of breeding
YY	Yellow seeds	Homozygous dominant	True-breeding for yellow seeds
Yy	Yellow seeds	Heterozygous	Hybrid, which does not breed true
yy	Green seeds	Homozygous recessive	True-breeding for green seeds

Mendel carried out a classic monohybrid cross, which has helped us to understand the inheritance of a single pair of contrasting alleles in garden peas.

Summary of Mendel's monohybrid cross experiment

Parents (Mendel selected these from greenhouse stock)

Phenotype: True breeding with   ×   True breeding with
            yellow seeds   green seeds

Genotype: **YY** (homozygous)   **yy** (homozygous)

Gametes: All contain **Y**   All contain **y**

Fertilisation: **Y** gametes fertilise **y** gametes and so all offspring are **Yy**

Offspring (F_1)

Genotype: All **Yy** (heterozygous)

Phenotype: All hybrids with yellow seeds because **Y** is dominant to **y**

Parents (Mendel crossed the F$_1$ generation by pollinating individuals with their own pollen)

Phenotypes: Hybrid with yellow seeds   ×   Hybrid with yellow seeds

Genotypes: **Yy** (heterozygous)   **Yy** (heterozygous)

Gametes: Both **Y** and **y**   Both **Y** and **y**

Fertilisation (various fertilisations are possible in the proportions shown in the following Punnett square diagram)

Gametes	Y	y
Y	YY	Yy
y	Yy	yy

Hints & tips ⭐

Punnett square diagrams can be used to work out offspring genotypes and phenotypes in their predicted proportions.

Offspring (F$_2$)

Phenotypes: 75% yellow seeds (**YY** and **Yy**) and 25% green seeds (**yy**)

This can be expressed as a ratio of 3 plants with yellow seeds : 1 plant with green seeds.

An example of human inheritance: cystic fibrosis (CF) in a family

Cystic fibrosis (CF) is caused when an individual inherits two copies of a recessive allele **c**. In the imaginary **family tree** in Figure 2.20, individuals 2, 3 and 4 are heterozygous for this condition. Each has the genotype **Cc**, which means that they do not have the condition but could each pass the **c** allele on to offspring – they are carriers. Individual 2 has passed **c** to her daughter 3 whose partner, 4, is also a carrier. In situations like this one, individuals 3 and 4 might receive genetic counselling before starting a family. This could help inform them of the chances of their offspring inheriting the condition and the implications involved. Can you work out the chances of their children inheriting CF?

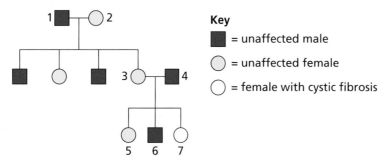

Key

◼ = unaffected male

○ = unaffected female

○ = female with cystic fibrosis

Figure 2.20 Cystic fibrosis in a family tree

Predicted and observed ratios

Offspring ratios can be predicted in advance of carrying out a cross because there is a statistical expectation of the result. However, observed ratios seen after carrying out crosses are usually slightly different from predicted ones. This is mainly because there are random processes, such as the fusion of gametes in fertilisation, involved. Chance plays a part, a bit like flipping a coin – you can predict a 1:1 ratio of heads to tails but this is not a guaranteed result! In the cystic fibrosis example above, it is possible to predict outcomes but these are not guaranteed.

Key words

Allele – a form of a gene

Continuous variation – variation which shows a range of values between extremes

Discrete – describes variation that is clear-cut and can be categorised into distinct groups

Dominant – form of a gene that is expressed in the phenotype, whether homozygous or heterozygous

F_1 – first generation of a monohybrid cross

F_2 – second generation of a monohybrid cross

Family tree – diagram that shows the inheritance of a genetic condition in a family

Genotype – the alleles that an organism has for a particular characteristic, usually written as symbols

Heterozygous – describes a genotype in which the two alleles for the characteristic are different

Homozygous – describes a genotype in which the two alleles for the characteristic are the same

Monohybrid cross – cross set up to study the inheritance of contrasting characteristics caused by alleles of the same gene

Parental generation (P) – parents of a monohybrid cross

Phenotype – the visible characteristics of an organism that occur as a result of its genes

Polygenic – inheritance determined by the interaction of several genes acting together

Punnett square diagram – a grid diagram which shows the predicted genotypes and proportions of offspring from a cross

Recessive – allele of a gene that only shows in the phenotype if the genotype is homozygous for that allele

Single gene inheritance – discrete inheritance patterns produced by alleles of one gene

Variation – differences in characteristics that can be seen between individual members of a species

Questions ?

A Short-answer questions

1 Give the term used to describe the differences that exist between organisms of the same species. (1)
2 State what is meant by the following terms:
   a) continuous variation
   b) discrete variation. (2)
3 Give an example of a characteristic in humans that shows continuous variation. (1)
4 Give an example of a characteristic in humans that shows discrete variation. (1)
5 State what is meant by the term 'single gene inheritance'. (1)
6 State what is meant by the term 'polygenic'. (1)
7 Describe the difference between the terms 'gene' and 'allele'. (2)
8 Describe the difference between the terms 'genotype' and 'phenotype'. (2)
9 Describe the difference between the terms 'homozygous' and 'heterozygous'. (2)
10 Describe the difference between the terms 'dominant' and 'recessive'. (2)
11 Give the genotypes of individuals 4 and 7 in the family tree in Figure 2.20 on page 58. (2)
12 In the human cystic fibrosis example on page 58, give the chance that a child of two heterozygous parents would inherit the condition. (1)
13 Explain why predicted ratios from a monohybrid cross are not always achieved. (2)

B Longer-answer questions

1 Using information from Key Area 3.6, describe the importance of variation to a species. (2)
2 In fruit flies, grey body is determined by the allele G and black body is determined by the allele g. Use this information to draw out a cross to show the inheritance of body colour in fruit flies. Show the P and F_1 phenotypes and genotypes of a cross between two heterozygous flies. (4)

Answers on page 77

Key Area 2.5
Transport systems: plants

Key points !

1 Plants require water for transporting materials such as **minerals**, for photosynthesis and for the maintenance of turgidity. ☐
2 Plants take up water by osmosis and minerals from the soil through the **root epidermis**. ☐
3 **Root hair cells** increase the surface area of the root epidermis. ☐
4 Water and minerals taken up by roots are passed into tubes made up of dead **xylem cells**. ☐
5 The cell walls of xylem cells contain a tough woody substance called **lignin**. ☐
6 Lignin allows the xylem vessels to withstand the pressure changes as water moves through the plant. ☐
7 Water and minerals pass through the plant stems to the leaves through xylem vessels. ☐
8 Water enters leaves in xylem vessels and passes into **spongy mesophyll** cells by osmosis. ☐
9 Water evaporates into spaces in the spongy mesophyll. ☐
10 Water vapour diffuses out of leaves through pores in the **leaf epidermis** called **stomata**. ☐
11 The stomata open and close by movements of **guard cells** found on either side of each pore. ☐
12 **Transpiration** is the process of water moving through a plant and its evaporation through the stomata. ☐
13 **Transpiration rate** is affected by wind speed, humidity, temperature and plant surface area. ☐
14 Sugars made by photosynthesis are transported up and down plants in **phloem** tubes, which are made from living cells. ☐
15 **Sieve tube cells** in phloem tissue have lost their nuclei and have end walls called sieve plates containing pores. ☐
16 Sieve tube cells in phloem tissue are associated with **companion cells** containing nuclei. ☐
17 Leaves have **veins** which contain xylem and phloem to carry water into leaves and sugars out of leaves. ☐
18 Roots, stems and leaves are plant organs. ☐
19 Most photosynthesis occurs in the closely packed green cells of the **palisade mesophyll**. ☐

Summary notes
The need for water

Plants require water because it is a raw material for photosynthesis, providing the hydrogen to convert carbon dioxide gas to sugar. The sugar

Hints & tips ★

There is more about photosynthesis in Key Area 3.3 on page 99.

produced by photosynthesis and minerals absorbed from the soil are transported around the plant dissolved in water. Water is also required for the maintenance of turgor in cells giving the plant support. It can also help to cool plants when it evaporates from leaves – see the table below.

Hints & tips

There is more about turgidity in Key Area 1.2 on page 5.

Use of water	Note
Photosynthesis	Hydrogen from water combines with carbon dioxide to produce sugar
Transport	Minerals and sugars are transported dissolved in water
Support	Water swells vacuoles making cells turgid
Cooling	Evaporation of water from leaves has a cooling effect

Plant transport systems

Water

Plants have a transport system for carrying water and minerals. Vast amounts of water pass through plants but only 1% of this is used by the plant cells for photosynthesis and turgor. The remaining 99% evaporates from the leaves and is lost to the atmosphere by the process of transpiration.

Movement of water into plants occurs by osmosis into the root hairs. Water and minerals are transported into xylem vessels, then up through the plant roots and stem. Xylem cells are dead and their walls contain lignin to withstand changes in pressure when water is moving through the plant. The lignified xylem also gives plant stems support and the strength to withstand wind.

Xylem carries water and minerals into leaves. Some water is used in photosynthesis but most evaporates out of stomata in the process of transpiration. Each stoma is formed between a pair of guard cells, whose action can open or close the pore. Figure 2.21(a) shows the main tissues in a leaf. Movement of water in a plant is summarised in Figure 2.21(b).

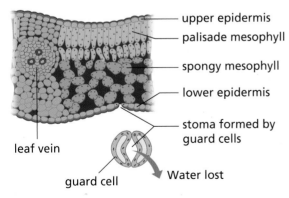

Figure 2.21(a) A magnified section through a leaf, showing the surface view of a stoma

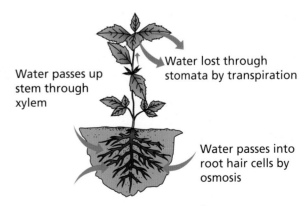

Figure 2.21(b) Water movement in a plant

Transpiration rate

Transpiration rate is the speed of movement of water through a plant and water loss from leaves. This rate can be expressed as volume of water lost per unit time per unit of leaf area. This rate is affected by various abiotic factors in the atmosphere including wind speed, humidity and temperature. Transpiration rate can be measured using a piece of apparatus called a potometer. Types of potometer include weight potometers and bubble potometers. An example of a bubble potometer is shown in Figure 2.22(a). The method shown in Figure 2.22(b) can be used to measure the unit areas of the leaves on the shoot.

> **Hints & tips**
>
> There is a diagram of a weight potometer on page 134.

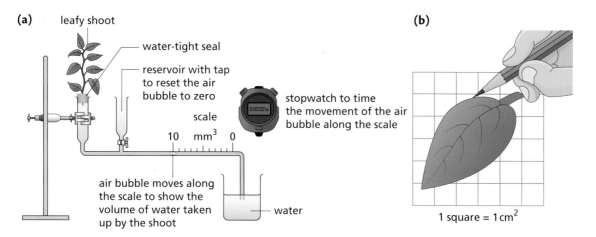

Figure 2.22 (a) Potometer used to measure the rate of water uptake by a leafy shoot which is taken as the rate of transpiration. **(b)** Method of measuring the surface area of a leaf

Use of a **potometer** is a technique you need to know for your exam.

> **Hints & tips**
>
> You should now try Q3 in the Skills of Scientific Inquiry section on page 133 which covers this technique.

The following table shows the effects of raising the levels of certain factors on the rate of transpiration.

Atmospheric factor	Effect of raising the level of factor on the rate of transpiration
Wind speed	Increase
Humidity	Decrease
Temperature	Increase

The surface area of the parts of the plant in contact with the air can also affect transpiration. Plants living in habitats with little free water have leaves with smaller surface areas; this is an adaptation to reduce transpiration and so help the plant retain the little water that it can find. This explains why in some desert plants, such as cacti, the leaves have been reduced to needles or spines.

Sugars

Sugars made in photosynthesis are transported around the plant in phloem. Phloem cells are living cells and able to transport sugars from photosynthetic tissues to all parts of the plant where they are needed for respiration. Phloem tissue usually runs alongside xylem tissues within the organs of the plant. Figure 2.23 shows the locations of xylem and phloem in a plant and gives details of the structure of these tissues.

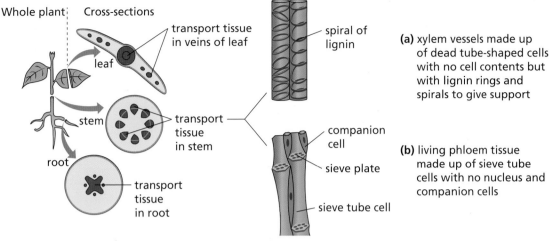

Figure 2.23 Distribution of transport tissues in a plant with detail of (a) xylem and (b) phloem

Key words

Companion cell – living phloem cell with a nucleus found associated with sieve tube cells

Guard cells – found on either side of a stoma; they control gas exchange in leaves by controlling opening and closing of the stoma

Leaf epidermis – skin-like tissue forming the upper and lower covering of leaves

Lignin – carbohydrate material lining the xylem vessels and providing strength and support

Minerals – nutrient ions essential for healthy growth

Palisade mesophyll – main photosynthetic tissue in leaf containing many chloroplasts

Phloem – tissue in plants that transports sugars

Potometer – apparatus used to measure transpiration rate in leafy shoots and plants

Root epidermis – outer layer of cells of a root

Root hair cell – specialised cell that increases the surface area of the root epidermis to improve the uptake of water and minerals

Sieve tube cell – phloem cell without a nucleus and with sieve plates at the ends

Spongy mesophyll – plant leaf tissue with loosely packed cells and air spaces between them to allow gas exchange

Stomata – tiny pores in the leaf epidermis that allow gas exchange (*sing.* stoma)

Transpiration – evaporation of water through the stomata of leaves

Transpiration rate – speed of water loss through leaves measured in volume of water per unit area of leaf per minute

Vein – tissue consisting of xylem vessels and phloem tubes running through a leaf

Xylem vessels – narrow, dead tubes with lignin in their walls for the transport of water and minerals in plants

Questions ?

A Short-answer questions

1 Give **one** reason why plants need a transport system. (1)
2 Give **two** examples of substances that are transported around a plant. (2)
3 State the part played by xylem vessels in plant transport. (1)
4 Name the process by which water vapour evaporates through the stomata of leaves. (1)
5 State the role of the guard cells in the epidermis of a leaf. (1)
6 State the role of phloem tubes in the transport system of a plant. (1)
7 Name **one** abiotic factor which can affect the rate of transpiration. (1)
8 Name the apparatus used to measure the transpiration rate from a leafy shoot. (1)

⇨

⇒
B Longer-answer questions

1 Explain why plants require water.   (2)
2 Name the tissues present in the structure of a plant leaf.   (4)
3 Describe the structures and processes involved in the uptake and movement of water from soil through a plant to the air.   (4)
4 Describe how you would use a bubble potometer (Figure 2.22 on page 62) to measure the rate of transpiration in a leafy shoot. **T**   (2)

Answers on page 77

Key Area 2.6
Transport systems: animals

Key points !

1 In mammals nutrients, oxygen and carbon dioxide are transported in the blood. ☐
2 Blood is pumped by a four-chambered **heart**. ☐
3 Blood is pumped from the heart to the lungs then back to the heart before being pumped to all other parts of the body. ☐
4 Two chambers called the left and right **ventricles** pump blood out of the heart and into **arteries**. ☐
5 The arteries carrying blood away from the heart to the lungs are the **pulmonary arteries**. ☐
6 The artery carrying blood away from the heart and which branches to all other parts of the body is the **aorta**. ☐
7 Branches of the aorta that supply the heart muscle itself with blood are the **coronary arteries**. ☐
8 Two chambers called the left and right **atria** receive blood into the heart from **veins**. ☐
9 Veins bringing blood to the heart from the lungs are the **pulmonary veins**. ☐
10 Veins bringing blood to the heart from all parts of the body join to form the **vena cava** before entering the heart. ☐
11 Arteries carry blood away from the heart and have thick, muscular walls and a relatively narrow internal diameter. ☐
12 Blood in arteries is under high pressure. ☐
13 Veins carry blood back to the heart and have thinner walls and relatively wider internal diameters than similar-sized arteries. ☐
14 Blood in veins is under lower pressure than blood in the arteries. ☐
15 To prevent the low-pressure blood from flowing backwards, veins have **valves**. ☐
16 Within organs and tissues, blood is carried in networks of **capillaries**. Capillaries link the arteries supplying the organ or tissue with the veins, which carry the blood away from the organ or tissue. ☐
17 Capillaries have thin walls and their networks produce a large surface area through which material can be exchanged with tissues efficiently. ☐
18 In mammals, the blood contains liquid called **plasma** along with **red blood cells** and **white blood cells**. ☐
19 Red blood cells are specialised because they are a **biconcave** shape, which increases their surface area, and have **haemoglobin**, a red oxygen-carrying pigment. ☐
20 In the lungs, oxygen combines with haemoglobin in the blood to form **oxyhaemoglobin** which carries the oxygen round the body. ☐
21 White blood cells are parts of the **immune system** and are involved in destroying **pathogens**, which are disease-causing micro-organisms such as certain bacteria, viruses and fungi. ☐
22 **Phagocytes** are white blood cells which carry out phagocytosis by engulfing and digesting pathogens. ☐
23 **Lymphocytes** are also white blood cells. Some of these cells produce antibodies which destroy pathogens. ☐
24 Each type of antibody is specific to a particular pathogen. ☐

Summary notes
Animal transport systems

In mammals, oxygen, carbon dioxide, nutrients (such as glucose), wastes and other substances are transported in the blood as it circulates round their bodies. This is summarised in the following table.

Substance transported	Transport route
Nutrients	From small intestine to all cells
Nitrogenous wastes	From all cells to kidneys
Oxygen	From lungs to all cells
Carbon dioxide	From all cells to lungs
Hormones	From glands to target organs

Pathway of blood in circulation

The circulatory system consists of the heart, blood vessels and the blood they contain.

The human circulatory system is a double circulation where the blood flows twice through the heart for each complete circuit. The lungs have a separate circulation from the rest of the body, which increases the efficiency because oxygenated blood can be kept separate from deoxygenated blood. Figure 2.24 shows the pathway of blood through the heart, lungs and body.

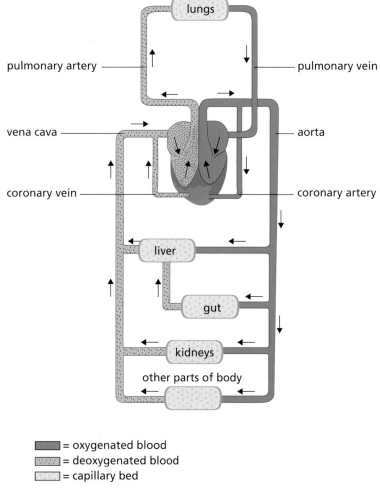

= oxygenated blood
= deoxygenated blood
= capillary bed

Figure 2.24 Pathway of blood through the heart, lungs and body

Hints & tips

A for Artery, A for Away
V for Veins, V for Valves;
veIN to the heart
CaPillArieS let substances
PASs through their thin
walls

Blood vessels

There are three main types of blood vessel in the circulatory system of a mammal, each with distinct features and functions. This is summarised in the table below.

Type of blood vessel	Function	Features
Arteries relatively narrow central channel thick muscular wall	Carry blood away from the heart	Thick muscular wall Pulse can be detected Blood at high pressure
Veins wider central channel valve thinner muscular walls	Carry blood towards the heart	Thinner muscular wall No pulse detected Blood at low pressure Contain valves
Capillaries thin wall (one cell thick)	Carry blood through tissues and organs Exchange of materials	Thin wall, one cell thick

Circulation

When an artery from the heart reaches an organ it divides up into a network of very narrow capillaries. The capillaries join up into a vein to carry blood out of the organ and back to the heart. The capillaries within an organ create a large surface area in contact with the cells of the organ. This allows efficient exchange of materials between the cells of the organ and the bloodstream.

Heart structure

The heart is a muscular pump, which has four chambers. There are two atria (top chambers of the heart), the right atrium and the left atrium, and two ventricles (bottom chambers of the heart), the right ventricle and the left ventricle. The heart has four valves, which prevent the backflow of blood, so that blood flows in one direction only.

Major blood vessels carry blood into and out of the heart. The main veins return blood under low pressure and the main arteries take blood under high pressure out of the heart.

Hints & tips

Arteries and veins are like motorways between cities carrying quantities of goods and transporting waste materials. The capillaries are like the busy streets within the city where loading and unloading of the goods and waste materials take place.

Pathway of blood through the heart

Deoxygenated blood from the body returns to the heart in the vena cava. It enters the right atrium and is then pumped through a valve into the right ventricle. The right ventricle then pumps the deoxygenated blood through another valve and out through the pulmonary artery to the lungs where it picks up oxygen.

Oxygenated blood from the lungs returns to the heart in the pulmonary vein. It enters the left atrium and is pumped through a valve into the left ventricle. The thick cardiac muscle of the left ventricle then pumps the oxygenated blood through another valve and out through the aorta to the body. Figure 2.25 shows a section through the heart with its four valves, its associated blood vessels and the pathway of blood through it.

□ deoxygenated blood □ oxygenated blood

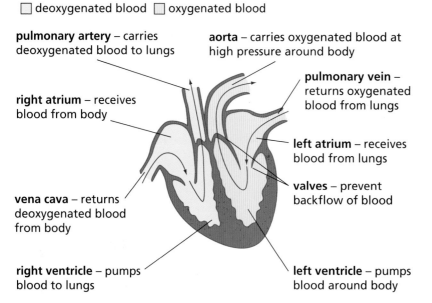

pulmonary artery – carries deoxygenated blood to lungs

aorta – carries oxygenated blood at high pressure around body

pulmonary vein – returns oxygenated blood from lungs

right atrium – receives blood from body

left atrium – receives blood from lungs

vena cava – returns deoxygenated blood from body

valves – prevent backflow of blood

right ventricle – pumps blood to lungs

left ventricle – pumps blood around body

Figure 2.25 Structure of the heart and its associated blood vessels

The coronary blood vessels

The heart muscle requires its own blood supply to provide it with oxygen and glucose for respiration. This is provided by the coronary blood vessels, which can be seen on the outside of the heart in Figure 2.26. The coronary arteries branch off the aorta and carry oxygenated blood directly to the heart muscle. The coronary veins carry the deoxygenated blood away and connect to the vena cava. Heart disease results if the coronary arteries become narrow or blocked. A heart attack occurs when a coronary vessel becomes blocked. The region of the heart muscle it supplies is starved of oxygen and dies. If only a small area of heart muscle dies, a mild heart attack occurs and the person may recover fully. If a large area of heart muscle dies, a severe heart attack occurs and the person may die.

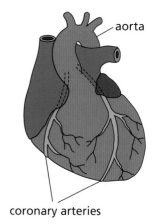

aorta

coronary arteries

Figure 2.26 Coronary blood supply

Pulse

Each time the heart muscle contracts, a small volume of blood is pumped at high pressure into the arteries. This produces a wave of pressure that can be felt as a pulse beat in an artery, especially where the artery comes near to the surface of the body, for example at the wrist.

Blood

Blood is composed mainly of plasma. This is a watery, yellow liquid in which blood cells float. There are several types of cell in blood including red blood cells (Figure 2.27), which give the blood its colour, and white blood cells (Figure 2.28), which are involved in destroying pathogens as part of the immune system.

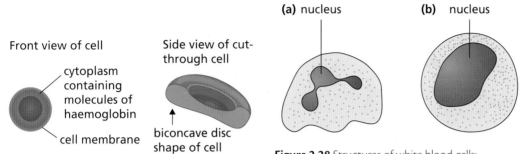

Figure 2.27 Structure of a red blood cell

Figure 2.28 Structures of white blood cells: **(a)** phagocyte and **(b)** lymphocyte

Red blood cells

Red blood cells are very small and numerous and their cytoplasm contains the red pigment haemoglobin. They have no nucleus, which gives more space for the haemoglobin, and are flattened into a biconcave disc shape to increase surface area for gas exchange.

The transport of oxygen

Red blood cells are specialised to carry oxygen attached to a protein called haemoglobin. Haemoglobin can combine rapidly with oxygen in the highly oxygenated environment of the lungs. Haemoglobin can then release oxygen rapidly in the poorly oxygenated environment of other tissues. This is shown in the following equation:

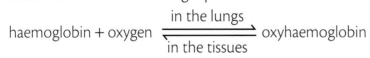

> **Hints & tips** ★
>
> *You must be familiar with this equation for your exam.*

White blood cells

Pathogens are disease-causing micro-organisms such as certain bacteria, viruses and fungi. Phagocytes are white blood cells which can engulf and digest pathogens in a process called phagocytosis as shown in Figure 2.29.

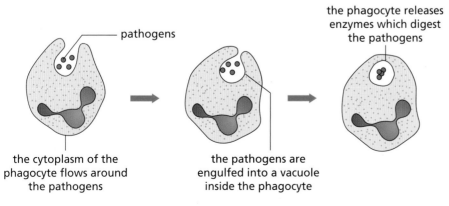

Figure 2.29 Stages of phagocytosis

Some lymphocytes produce antibodies which are specific to a particular pathogen as shown in Figure 2.30. The antibodies have a complementary shape to the pathogen, allowing them to bind to the pathogen and destroy it.

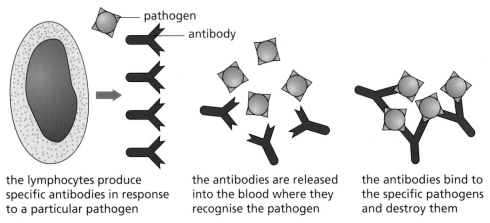

the lymphocytes produce specific antibodies in response to a particular pathogen

the antibodies are released into the blood where they recognise the pathogen

the antibodies bind to the specific pathogens and destroy them

Figure 2.30 The specific action of antibodies

Key words

Aorta – main artery that carries oxygenated blood away from the heart in mammals

Artery – general name for a blood vessel that carries blood away from the heart

Atria – upper chambers of the heart, which receive blood from veins (*sing.* atrium)

Biconcave – shape of a red blood cell describing dimples which increase the surface area of the cell

Capillaries – tiny blood vessels with walls one cell thick where exchange of materials occurs

Coronary arteries – blood vessels that carry oxygenated blood to the heart tissues

Haemoglobin – pigment in red blood cells that transports oxygen as oxyhaemoglobin

Heart – muscular organ that pumps blood around the body

Heart valves – structures in the heart which prevent backflow of blood

Immune system – organs and tissues which resist infections

Lymphocytes – white blood cells, some of which can produce antibodies to destroy specific pathogens

Oxyhaemoglobin – substance produced when haemoglobin combines with oxygen in lungs

Pathogen – a disease-causing organism such as certain bacteria, viruses and fungi

Phagocyte – white blood cell which can engulf and digest pathogens during phagocytosis

Plasma – liquid part of blood in which cells are suspended

Pulmonary artery – artery carrying deoxygenated blood from the heart to the lungs

Pulmonary vein – vein carrying oxygenated blood to the heart from the lungs

Red blood cell – blood cell containing the pigment haemoglobin responsible for the transport of oxygen

Valve – structure that prevents the backflow of blood within the heart and in veins

Vein – general name for a blood vessel with valves that transports blood to the heart

Vena cava – vein carrying deoxygenated blood to the heart from the body systems

Ventricles – lower chambers of the heart that receive blood from the atria and pump it into arteries

White blood cells – phagocytes and lymphocytes

Questions ?

A Short-answer questions

1 Describe the function of the heart. (1)
2 Describe the general functions of the following blood vessel types:
   a) arteries
   b) veins. (2)
3 Name the veins which carry out the following functions:
   a) return deoxygenated blood from the body to the heart
   b) transport oxygenated blood from the lungs to the heart. (2)
4 State the general role of valves. (1)
5 Name the artery that supplies the cardiac muscle of the heart with blood. (1)
6 Name the arteries which carry out the following functions:
   a) supply the body systems with blood from the heart
   b) supply the lungs with blood from the heart. (2)
7 Name the red-coloured substance which carries oxygen in blood. (1)
8 Name the process by which certain white blood cells engulf pathogens. (1)
9 Name the substances released by certain lymphocytes which destroy specific pathogens. (1)
10 Use the information in Figure 2.24 (on page 66) to account for the difference between the thickness of the muscular wall of the left ventricle and the wall of the right ventricle. (2)

B Longer-answer questions

1 Explain how the structure of red blood cells helps them carry out their function. (3)
2 Describe the process of phagocytosis. (3)
3 Explain the meaning of the term 'specific' with reference to the action of antibodies. (3)
4 Describe the circulation of blood through the heart. (5)

Answers on page 78

Absorption of materials

Key points

1 Oxygen and nutrients from food are absorbed into the bloodstream and delivered to cells for respiration. ☐

2 Waste materials, such as carbon dioxide, are removed from cells into the bloodstream. ☐

3 Tissues contain **capillary networks** to allow the exchange of materials at cellular level. ☐

4 Surfaces involved in the **absorption** of materials have large surface area, thin walls and extensive blood supply to increase the efficiency of absorption. ☐

5 **Lungs** are gas exchange organs consisting of a large number of thin-walled sacs called **alveoli**. ☐

6 Alveoli form a large surface area and have an extensive blood supply for efficient gas exchange. ☐

7 Oxygen diffuses from inhaled air in the lungs, through the alveolar walls and into the blood capillaries. ☐

8 Carbon dioxide diffuses from the blood capillaries and enters the alveoli from where it is breathed out. ☐

9 The **small intestine** is the organ involved in the absorption of nutrients from food. ☐

10 The large number of thin-walled **villi** provides a large surface area. ☐

11 Each villus contains a network of blood capillaries to absorb glucose and amino acids from food. ☐

12 Each villus contains a **lacteal** to absorb **fatty acids and glycerol** from food. ☐

Summary notes
Gas exchange
The lungs

The lungs (Figure 2.31) are the organs in which the gases oxygen and carbon dioxide are exchanged between the blood and the air.

Inhaled air passes into the lungs through breathing tubes and finally reaches the tiny alveoli, which make up the large gas exchange surface. Oxygen diffuses from inhaled air through the thin walls of the alveoli and into their extensive blood capillary network.

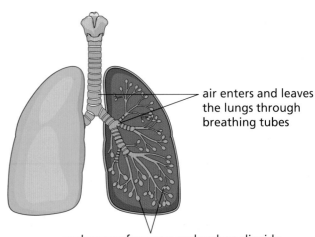

air enters and leaves the lungs through breathing tubes

exchange of oxygen and carbon dioxide occurs in the alveoli which are found at the ends of the breathing tubes

Figure 2.31 The structure of human lungs

Once in the blood, the oxygen passes into red blood cells. Carbon dioxide diffuses out of the blood capillaries and into the air in the alveoli (Figure 2.32). Exhaled air passes out of the lungs through the breathing tubes.

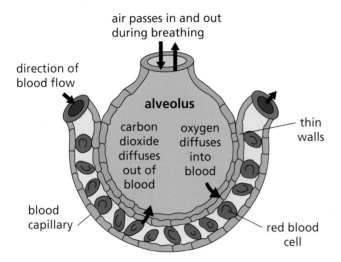

Figure 2.32 Gas exchange at an alveolus

Efficiency of the gas exchange surface

The internal structure of the lungs has many features that promote diffusion of gases and make gas exchange efficient. These are outlined in the following table.

Feature	Provided by...
Large surface area	Many alveoli in contact with an extensive capillary network
Extensive blood supply	Highly branched capillary network around each alveolus
Very thin walls	The single-cell thick walls of the alveoli and capillaries

Absorption of nutrients
The need for digestion

Many of the large carbohydrate, fat and protein molecules present in the food that we eat are insoluble. **Digestion** involves the breakdown of large, insoluble food molecules into small, soluble molecules by enzymes. This is required because only soluble molecules can dissolve and diffuse through the wall of the small intestine and enter the blood.

The small intestine

Soluble molecules produced by digestion pass through the small intestine wall (Figure 2.33) by diffusion and enter the blood capillaries and the lacteals.

The structure of the small intestine allows very efficient absorption to take place. It is long and the inner lining is folded and has many finger-like projections called villi (Figure 2.34), which increase the surface area available for the absorption of the soluble food molecules.

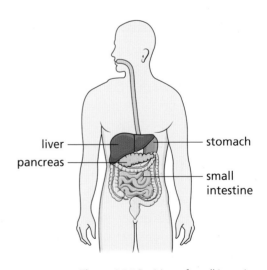

Figure 2.33 Position of small intestine in the body

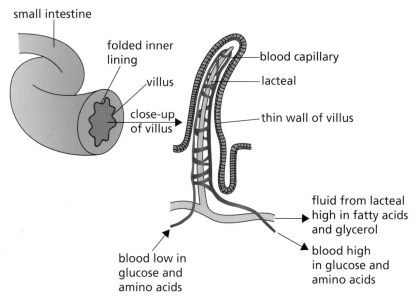

Figure 2.34 Villi lining the small intestine and structure of a single villus

The features of the villi that promote efficient absorption are outlined in the table below.

Feature of the villi	Advantage
Large surface area	Many villi in contact with an extensive capillary network
Extensive blood supply	Highly branched capillary network with a large surface area into which glucose and amino acid molecules are absorbed
Central lacteal	Fine tube gives a large surface area into which fatty acid and glycerol molecules are absorbed
Very thin walls	The single-cell thick walls of the villi promote efficient absorption

Key words

Absorption – the taking up of molecules by cells
Alveoli – tiny sacs in lungs that form the gas exchange surface (*sing.* alveolus)
Capillary network – branching system of tiny blood vessels providing a large surface area for absorption
Digestion – breakdown of large, insoluble food molecules into smaller, soluble ones
Fatty acids – products of the digestion of fat absorbed into lacteals
Glycerol – product of fat digestion absorbed into lacteals
Lacteal – central vessel in the villi responsible for the absorption of the products of fat digestion
Lungs – organs responsible for gas exchange
Small intestine – part of the digestive system where most absorption of nutrients occurs
Villi – finger-like projections of the small intestine lining providing a large surface area for absorption of nutrients from food (*sing.* villus)

Questions ❓

A Short-answer questions

1 Give **two** features of a surface involved in absorption. (2)
2 Name the sacs within lungs through which gases are exchanged. (1)
3 Name the finger-like projections that line the inside of the small intestine. (1)
4 Name the food molecules absorbed by the blood capillaries in villi. (1)
5 Name the food molecules absorbed by the lacteals in villi. (1)

B Longer-answer questions

1 Explain why oxygen and nutrients must be delivered to cells. (2)
2 Describe the features which surfaces involved in the absorption of materials have in common. (3)
3 Describe the features of a capillary network which increase its efficiency in absorption. (2)
4 Describe the features of an alveolus which make it efficient at exchanging gases. (3)
5 Describe the features of a villus which make it efficient at absorbing nutrients from food. (4)
6 Describe the similarities in the structure and function of the alveoli and the villi. (4)

Answers on page 78

Answers

Key Area 2.1

A Short-answer questions

1 tissue (1)
2 a) mesophyll, epidermis, xylem, phloem (other possible answers) [any 1] (1)
   b) blood, muscle, nervous (other possible answers) [any 1] (1)
3 group of tissues working together (1)
4 a) leaf, stem, root, flower, fruit (other possible answers) [any 1] (1)
   b) brain, heart, kidney (other possible answers) [any 1] (1)
5 group of organs working together (1)
6 nervous, circulatory, respiratory, digestive, excretory, reproductive, skeletal (other possible answers) [any 2] (2)
7 diploid (1)
8 chromatids (1)
9 attachment of chromatids; separation of chromatids [1 each] (2)
10 modified/developed for a specific function/differentiated (1)
11 can self-renew/divide; can specialise/differentiate [1 each] (2)
12 red blood cell/lymphocyte/phagocyte/neuron/muscle cell (other possible answers) [any 2] (2)
13 their DNA replicates; and chromatids are formed [1 each] (2)
14 so that daughter cells are genetically identical to the parent cells/have all the genetic information needed to carry out all their activities (1)
15 they have the ability to differentiate; and become specialised for a specific function [1 each] (2)

B Longer-answer questions

1 basic unit is the cell; similar cells form tissues; groups of tissues make up organs; organs working together make up systems [1 each] (4)
2 chromosomes replicate to form pairs of chromatids; spindle forms; chromosomes line up on the equator of the spindle; pairs of chromatids move apart [1 each] (4)
3 provides new cells for growth of organism; provides new cells for repair of damaged parts; maintains the diploid chromosome complement [1 each] (3)
4 loses its nucleus; develops a biconcave shape; produces haemoglobin [1 each] (3)

Answers

Key Area 2.2a

A Short-answer questions

1 brain; spinal cord   [1 each]   (2)
2 sorts out stimuli/messages/processes electrical impulses; sends electrical impulses along neurons to effectors   [1 each]   (2)
3 as electrical impulses   (1)
4 synapses   (1)
5 convert stimuli into electrical impulses   (1)
6 protect the body   (1)

B Longer-answer questions

1 cerebrum is site of reasoning/conscious thought/memory/emotions; cerebellum controls balance/coordination; medulla controls breathing/heart rate   [1 each]   (3)

2 sensory neurons carry impulses from receptors to the CNS/inter neuron; inter neurons carry impulses from sensory neuron to the motor neuron; motor neurons carry impulses from the CNS/inter neuron to effectors   [1 each]   (3)
3 chemicals are released from a neuron; chemicals pass into/diffuse across synapse; impulse created in next neuron   [1 each]   (3)
4 receptor stimulated/affected by a stimulus; sensory neuron carries impulse/message into spinal cord; inter neuron passes impulse/message to motor neuron; motor neuron triggers response in a muscle/gland/effector [1 each]   (4)

Answers

Key Area 2.2b

A Short-answer questions

1 release hormones into the bloodstream   (1)
2 pancreas (other possible answers)   (1)
3 travel in the bloodstream/plasma   (1)
4 target tissues   (1)
5 insulin; glucagon   [1 each]   (2)

B Longer-answer questions

1 target tissue has receptors; which are complementary to a specific hormone; hormone binds to receptor; and triggers a specific response   [1 each]   (4)
2 pancreas detects increase in blood glucose concentration; pancreas releases insulin into bloodstream; insulin binds to receptors on target tissue in liver; liver converts glucose to glycogen   [1 each]   (4)

Answers

Key Area 2.3

A Short-answer questions

1 haploid: one set of chromosomes in the nucleus; diploid: two sets of chromosomes in the nucleus [1 each]   (2)
2 male: anthers; female: ovaries   [1 each]   (2)
3 male: pollen; female: ovules   [1 each]   (2)
4 male: testes; female: ovaries   [1 each]   (2)
5 male: sperm; female: egg cells/ova   [1 each]   (2)
6 zygote   (1)
7 sperm tiny with tail; egg cells larger with food [1 each]   (2)

B Longer-answer questions

1 body cell produced by mitosis which maintains the diploid chromosome complement; gametes must be haploid so that diploid number restored after fertilisation   [1 each]   (2)
2 anthers make pollen; pollen transferred to ovary; pollen nucleus fuses/joins with/fertilises ovule nucleus; an embryo formed [1 each]   (4)
3 testes make sperm; ovary makes egg cells/ova; sperm nucleus fertilises egg/ovum nucleus; an embryo formed   [1 each]   (4)

Answers

Key Area 2.4

A Short-answer questions

1 variation (1)

2 a) variation that shows a range of differences between a minimum and a maximum value (1)

   b) variation in which the values are clear-cut and can be put into distinct groups (1)

3 height, weight (other possible answers) (1)

4 blood group, eye colour (other possible answers) (1)

5 controlled by alleles of one gene (1)

6 controlled by the alleles of many genes (1)

7 a gene controls a characteristic; alleles are forms of a gene   [1 each] (2)

8 genotype is the alleles an organism has for a characteristic; phenotype is the appearance of the organism as a result of its alleles   [1 each] (2)

9 homozygous means having two identical alleles of the same gene; heterozygous means having two different alleles of the same gene [1 each] (2)

10 dominant alleles always show in the phenotype; recessive alleles only show if they are in the homozygous form   [1 each] (2)

11 **4** Cc; **7** cc   [1 each] (2)

12 1 chance in 4 or 25% chance   [either] (1)

13 fertilisation is a chance process; so very many offspring are needed to achieve the predicted ratio   [1 each] (2)

B Longer-answer questions

1 variation gives the differences on which natural selection acts; and allows survival of the fittest giving evolutionary change [1 each] (2)

2 P: Grey-bodied Gg crossed with grey-bodied Gg; gametes of each parent are G and g; F_1 genotypes can be GG, Gg and gg **OR** draw Punnett square; 3 grey-bodied flies for each black-bodied fly [1 each] (4)

Key Area 2.5

A Short-answer questions

1 transport water to leaves for photosynthesis **OR** transport food around plant (1)

2 sugar/water/minerals   [any 2] (2)

3 transport of water/minerals (1)

4 transpiration (1)

5 change shape to open and close the stomata to control water loss through transpiration (1)

6 transport sugars around the plant (1)

7 wind speed/temperature/humidity/light [any 1] (1)

8 potometer (1)

B Longer-answer questions

1 raw material for photosynthesis; provides support for plant cells/turgidity [1 each] (2)

2 epidermis; spongy mesophyll; palisade mesophyll; vein/phloem and xylem   [1 each] (4)

3 water taken into roots/root hair cells by osmosis; water passes up xylem; water drawn into spongy mesophyll of leaf by osmosis; water evaporates into air spaces in spongy mesophyll/leaf; water moves out of leaf to air through stomata/water lost by transpiration [any 4] (4)

4 measure the volume of water taken up/ distance bubble moves on scale; in a measured period of time/minute   [1 each] (2)

Answers

Key Area 2.6

A Short-answer questions

1. pumps blood around body (1)
2. a) carry blood away from the heart (1)
   b) carry blood towards the heart (1)
3. a) vena cava (1)
   b) pulmonary vein (1)
4. prevent backflow of blood (1)
5. coronary artery (1)
6. a) aorta (1)
   b) pulmonary artery (1)
7. haemoglobin (1)
8. phagocytosis (1)
9. antibodies (1)
10. left ventricle has to pump blood to the body which is further; than the right ventricle which pumps to the lungs   [1 each]   (2)

B Longer-answer questions

1. red blood cells have no nucleus which gives extra space inside; they are biconcave which increases their surface area for gas exchange; they contain haemoglobin which carries oxygen   [1 each]   (3)
2. phagocyte flows round a pathogen; pathogen engulfed by phagocyte; pathogen destroyed/digested by phagocyte   [1 each]   (3)
3. each pathogen has a specific shape; the antibody shape is complementary; the antibody binds to the pathogen and destroys it   [1 each]   (3)
4. blood enters right atrium (from body); passes from right atrium to right ventricle; right ventricle pumps blood out of heart (to lungs); blood enters left atrium from lungs; passes from left atrium to left ventricle; left ventricle pumps blood out of heart (to body); correct mention of valve(s)   [any 5, 1 mark each]   (5)

Answers

Key Area 2.7

A Short-answer questions

1. large surface area; thin walls; extensive blood supply   [any 2]   (2)
2. alveoli (1)
3. villi (1)
4. glucose and amino acids (1)
5. fatty acids and glycerol (1)

B Longer-answer questions

1. needed for respiration; to provide energy to cells   [1 each]   (2)
2. thin walls; large surface area; extensive blood supply   [1 each]   (3)
3. highly branched to give an extensive surface area; very thin walls/walls one cell thick   [1 each]   (2)
4. large surface area; thin walls; extensive blood supply/network of capillaries [1 each]   (3)
5. large surface area; thin walls; extensive blood supply/network of capillaries; has a lacteal   [1 each]   (4)
6. both have large surface area; thin walls; extensive blood supply/network of capillaries; both absorb material   [1 each]   (4)

Practice assessment: Multicellular Organisms (45 marks)

Section 1 (10 marks)

1 The diagram below shows a single neuron.

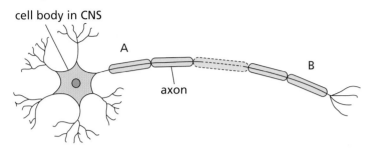

Which line in the table below correctly identifies this type of neuron and the direction of electrical impulse along its fibre?

	Type of neuron	Direction of impulse
A	Sensory	A → B
B	Motor	B → A
C	Sensory	B → A
D	Motor	A → B

2 The list shows terms used to describe inherited variation.

1 continuous
2 single-gene inheritance
3 polygenic
4 discrete

Which of the terms refers to variation in which there are clear-cut differences?

A 1 and 2
B 1 and 3
C 2 and 4
D 3 and 4

3 Which line in the table below correctly describes phloem cells?

	Description of cell	Substance(s) transported
A	Live	Sugars
B	Dead	Sugars
C	Live	Water and minerals
D	Dead	Water and minerals

4 Which line in the table below correctly describes arteries and veins?

	Arteries	Veins
A	Low pressure	Wide central channel
B	High pressure	Wide central channel
C	Low pressure	Narrow central channel
D	High pressure	Narrow central channel

5 Which structures form part of the lungs and contain gases which are exchanged between the air and the bloodstream?
   A villi
   B alveoli
   C capillaries
   D lacteals

6 The diagram below shows part of the human digestive system.

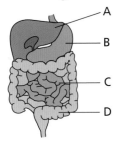

   In which structure would lacteals be found?

7 Which line in the table below matches white blood cells with their functions?

| | White blood cells | |
	Lymphocytes	Phagocytes
A	Produce antibodies	Absorb oxygen
B	Engulf pathogens	Produce antibodies
C	Absorb oxygen	Engulf pathogens
D	Produce antibodies	Engulf pathogens

Questions 8 and 9 refer to the graph below, which shows how increasing the percentage of carbon dioxide in the air affects the volume of air inhaled each minute.

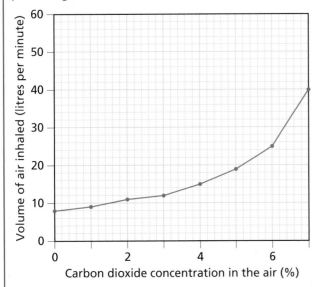

8 The volume of carbon dioxide entering the lungs each minute when the carbon dioxide concentration in the air is 7% is:
   A 2.8 litres   B 17.5 litres   C 28.0 litres   D 40.0 litres

9 When the carbon dioxide concentration of the air is increased from 2% to 5%, the volume of air inhaled increases by:
   A 6 litres per minute   B 8 litres per minute   C 10 litres per minute   D 19 litres per minute

10 The graph below shows how the percentage saturation of haemoglobin changes as the concentration of oxygen in the fluid surrounding it increases.

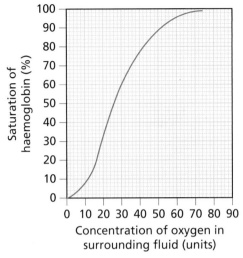

Which of the following conclusions can be drawn from the graph?

A As the oxygen concentration in the surrounding fluid increases, the percentage saturation of haemoglobin increases steadily.

B The highest percentage saturation of haemoglobin occurs when oxygen concentration is lowest.

C The rate of increase in percentage saturation of haemoglobin decreases at the highest oxygen concentrations.

D The lowest oxygen concentrations stimulate the most rapid increase in percentage saturation of haemoglobin.

Section 2 (35 marks)

1 The statements in the table below relate to the cells of a multicellular organism.
Copy the table and then decide if each statement is true or false and tick the appropriate box. If you decide that the statement is false write the correct word into the correction box to replace the word underlined in the statement. (3)

Statement	True	False	Correction
Similar cells with the same function are grouped into tissues in multicellular organisms			
Stem cells are specialised cells in animals that can divide and become different types of cell			
Mitosis maintains the haploid chromosome number			

2 The diagram shows part of the human central nervous system (CNS).

a) Name part X. (1)

b) Copy the table below, which shows the results of direct damage to various parts of the brain. Predict the part of the brain that is most likely to have been damaged in each case and add the information to the table. (3)

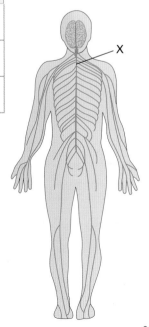

Part of brain damaged	Result
	Ability to balance impaired
	Loss of ability to control heart rate
	Reduced ability to remember simple experiences

3 The diagram below shows a section through a sweet pea flower.

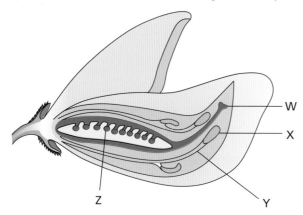

a) Identify the letter that shows the site of production of male gametes in this flower. (1)
b) Copy the sentence below, choosing the correct term from the choice in brackets to make it correct. (2)
The female gametes of a flowering plant are found within the (ovaries/stamens) and have the (diploid/haploid) number of chromosomes.

4 The diagram below shows part of a bubble potometer used to investigate the effect of increasing temperature on transpiration rate in a young plant seedling.

a) Copy and complete the table below, which refers to the location of various structures labelled on the diagram. (3)

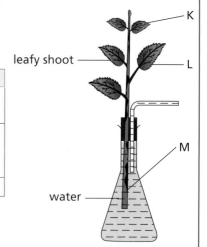

Letter	Name of structure	Function in transport
K		Allow water vapour to escape out of the leaf to the atmosphere
L		Have air spaces between cells into which water can evaporate before passing out of leaf
M	Xylem vessels	

b) The table below shows how the transpiration rate from the leaves of this plant varied as the temperature was raised during the experiment.

Temperature (°C)	Transpiration rate (cm³ water per hour per cm² leaf surface)
10	2.4
20	4.6
30	10.2
40	10.8

(i) On a piece of graph paper, plot a line graph of these data to show temperature against the transpiration rate. (2SSI)
(ii) Give **one** variable that must be kept constant at each temperature to ensure that a valid conclusion could be drawn using the results of this experiment. (1SSI)
(iii) Calculate the average rate of transpiration over the temperatures tested. (1SSI)
(iv) Using the data in the table, give the expected transpiration rate of this plant at 35°C. (1SSI)

5 The table below shows some terms involved in the description of variation and inheritance in living organisms. Copy and complete the table by adding the missing term and descriptions. (2)

Term	Description
Genotype	
	An organism with two identical alleles of a particular gene

6 The diagram on the right shows a plan of the basic circulatory system of a mammal.
   a) Name the chamber of the heart that pumps blood into vessel B. (1)
   b) Name blood vessel F. (1)
   c) Describe why vessels such as E have valves. (1)
   d) Give a letter on the circulation plan that shows where a capillary blood vessel could be found. (1)

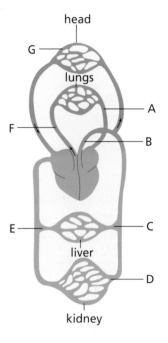

7 The diagram below represents undamaged and damaged alveoli in the human lungs.

undamaged

damaged

Identify a feature of the alveoli which has been affected by the damage and describe the effect this would have on gas exchange. (2)

8 Give an account of how a rise in blood glucose concentration in humans is controlled following its detection by receptors in the pancreas. (2)

9 The diagram on the right represents a cell during mitosis.
   a) Name the part labelled X. (1)
   b) Describe what would happen next during the process of mitosis. (1)

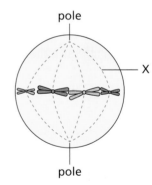

10 Read the following passage and answer the questions based on it.

Future uses of stem cells

When people imagine stem cell treatments in the future they often think of stem cell therapies such as transplants. Before a stem cell transplant, embryonic stem cells would be triggered to specialise into the required adult cell type. Then, those mature adult cells would be transplanted into a patient to replace tissue damaged by disease or injury. This type of treatment might be used to replace neurons damaged by spinal cord injury, stroke, Alzheimer's disease or Parkinson's disease. It could be used to produce insulin that could treat people with diabetes or to produce heart muscle cells that could repair damage after a heart attack.

Embryonic stem cell research could do much more. Studying how stem cells develop into heart muscle cells could provide clues about how we could encourage heart muscle to repair itself after a heart attack. Embryonic stem cells could be used to study the progress of disease, identify new drugs, or screen drugs for toxic side effects. Any of these would have a significant impact on human health without transplanting a single cell.

   a) Identify the source of the stem cells mentioned in the passage. (1)
   b) Name the type of cell which would be replaced in future treatment of stroke. (1)
   c) Identify the treatment mentioned in the passage which could be used for diabetes. (1)
   d) Give one example of how stem cells might be used in the development of drugs. (1)
   e) Describe the general difference between stem cell therapy and stem cell research. (1)

Answers to practice assessment: Multicellular Organisms

Section 1

1 D, 2 C, 3 A, 4 B, 5 B, 6 C, 7 D, 8 D, 9 B, 10 C

Section 2

1 true
  false – <u>non-specialised</u> or <u>unspecialised</u>
  false – <u>diploid</u>   [1 per correct line]   (3)
2 a) spinal cord   (1)
  b) cerebellum
     medulla
     cerebrum   [1 per correct line]   (3)
3 a) X   (1)
  b) ovaries, haploid   [1 each]   (2)
4 a) K stomata
     L spongy mesophyll
     M carry water and minerals up stem to
     leaves   [1 each]   (3)
  b) (i) scales and labels; correctly plotted and
         points connected using a ruler
         [1 each]   (2)
     (ii) light intensity **OR** air movements
         OR size of seedling **OR** humidity   (1)
     (iii) 7.0 cm³ per hr per cm²   (1)
     (iv) 10.4–10.5 cm³ per hr per cm²   (1)
5 the different alleles of a gene an individual
  has for a particular characteristic;
  homozygous   [1 per correct line]   (2)

6 a) left ventricle   (1)
  b) pulmonary artery   (1)
  c) prevent backflow of blood   (1)
  d) G   (1)
7 feature: reduced surface area; effect:
  reduced gas exchange/decreased
  efficiency of gas exchange   [1 each]   (2)
8 pancreas releases insulin; insulin causes
  conversion of excess glucose to
  glycogen in the liver   [1 each]   (2)
9 a) spindle fibres   (1)
  b) (spindle fibres contract and)
     chromatids/chromosomes are
     pulled apart/to opposite ends of
     the cell   (1)
10 a) embryos   (1)
   b) neurons   (1)
   c) production of insulin   (1)
   d) identify new drugs **OR** screen
      drugs for side effects   (1)
   e) in therapy, patients receive
      treatment with stem cells; in
      research, stem cells are studied
      (to develop future treatments)   (1)

Life on Earth

Ecosystems

Key points !

1 **Biodiversity** is a term which refers to the variety and relative abundance of **species** present in an **ecosystem**. ☐

2 An ecosystem consists of all the organisms (the **community**) living in a particular place (the **habitat**) and the non-living components with which the organisms interact. ☐

3 A **population** is the total number of organisms of the same species that live in a particular habitat. ☐

4 Interactions of species with other species in the community include **competition** and **predation**. ☐

5 **Producers** are green plants which produce their own food by photosynthesis. ☐

6 **Consumers** obtain energy by eating other organisms. **Herbivores** eat plants only, **carnivores** eat animals only and **omnivores** eat both animals and plants. ☐

7 Some carnivores are **predators** which hunt and kill **prey** organisms. ☐

8 A **niche** is the position occupied by an organism and the role it plays within its community. ☐

9 Niche relates to the resources an organism requires in its ecosystem, such as light or nutrient availability, and its interactions with other organisms in the community. ☐

10 An organism's niche involves competition and predation and the abiotic conditions it can tolerate such as temperature. ☐

11 A food chain diagram represents how energy flows from one organism to another during feeding. ☐

12 Food chains are linked together as food webs to show how organisms in an ecosystem are linked through feeding. ☐

13 Competition in ecosystems occurs when resources are in short supply and organisms struggle for the same resources. ☐

14 **Interspecific competition** occurs among individuals of different species for one or a few of the resources they require. ☐

15 **Intraspecific competition** occurs among individuals of the same species and is for all resources required. ☐

16 Intraspecific competition is therefore more intense than interspecific competition. ☐

Summary notes
Components of an ecosystem

An ecosystem is a natural biological unit in which organisms interact with their environment. An ecosystem consists of all the species (the community) living in a particular place (the habitat) whose non-living components affect the organisms present:

ecosystem = community + habitat

A community is all the populations of living species which occupy a habitat. Each population is the number of members of a species in the habitat. A habitat is the place where a community lives and the non-living factors which influence it.

A moorland ecosystem

A Scottish moorland community consists of plants such as heathers and grasses and animals such as insects, mammals and birds.

The moorland habitat consists of an area of acidic soil and rock covering a hillside and affected by climatic conditions such as temperature and rainfall.

Biodiversity

Biodiversity is the term used to describe the variety and relative abundance of species present in an ecosystem. The biodiversity of the moorland ecosystem is the number of different species present and the relative sizes of their populations.

Food chains

A food chain shows the feeding relationship between organisms. It shows how energy flows between individual species. A food chain always starts with a producer, usually a green plant which produces its own food by photosynthesis.

Various consumer levels are shown connected by arrows, which indicate the direction of energy flow. Consumers which eat producers are called herbivores and those which eat other consumers are called carnivores. Omnivores eat a mixed diet of producers and other consumers. Predators are carnivores that hunt, and prey species are food that predators hunt for.

An example of a simple food chain taken from the moorland ecosystem is shown in Figure 3.1. Note the descriptions of the species below which can be seen from the drawing of the food chain.

> **Hints & tips** ★
>
> There is more about photosynthesis in Key Area 3.3 on page 99.

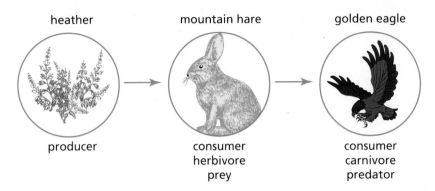

Figure 3.1 Moorland food chain

Food webs

A food web is made up of many interconnected food chains. A food web diagram shows the feeding relationships between members of a community. Figure 3.2 shows part of a food web for a moorland ecosystem. The species in the community are connected by arrows which indicate the direction of flow of energy between them as they feed.

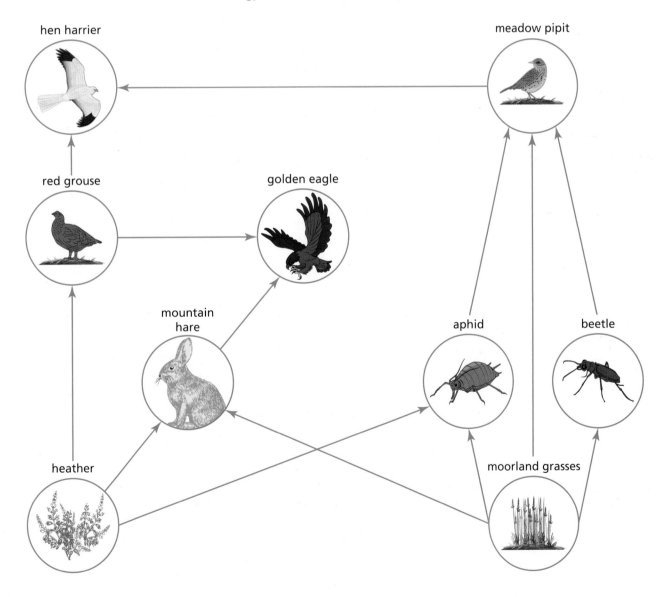

Figure 3.2 Part of a moorland food web

Niche

A niche is the role that an organism plays within a community. It relates to the resources it requires in its ecosystem, such as light and nutrient availability, and its interactions with other organisms in the community. It can involve competition and predation and the conditions that an organism can tolerate such as temperature, pH and moisture.

The following table shows some of the terms used to describe how an organism's niche relates to its place in the food web.

Term	Meaning in the food web	Example from moorland ecosystem
Producer	Green plant which makes food by photosynthesis	Heather
Consumer	Animal which obtains energy by eating other organisms	Red grouse
Herbivore	Consumer which eats plants only	Aphid
Carnivore	Consumer which eats other animals only	Golden eagle
Omnivore	Consumer which eats both plants and animals	Meadow pipit
Predator	Animal which obtains food by hunting and killing prey	Hen harrier
Prey	Animal which is hunted and eaten by a predator	Mountain hare

Competition in ecosystems

Competition in ecosystems occurs when resources such as food are in short supply.

- Intraspecific competition occurs among individuals of the same species and is for all resources required.
- Interspecific competition occurs among individuals of different species for one or a few of the resources they require.

Intraspecific competition is therefore more intense than interspecific competition.

Interspecific competition can be seen in food web diagrams. For example in Figure 3.1, red grouse and mountain hares are in competition because they both eat heather, and hen harriers and golden eagles are in competition because they both hunt and eat red grouse.

Effects of removal of species from food webs

Removal of any species from a food web has consequences for the population sizes of other species. Although it is impossible to predict exactly what will happen to populations when a species is removed, some general principles can be applied.

Hints & tips

Remember:

Intra – same species

Inter – between species

Removing the heather plants from the food web in Figure 3.1 could seriously reduce red grouse populations because they eat little else and would be short of food. The mountain hares which also eat heather may start to eat more moorland grass in the absence of heather thus reducing the grass population and possibly reducing other consumers such as aphids.

Removing a herbivore species, such as the beetles, from the web might cause increases in moorland grass numbers because they are eaten less but might decrease the meadow pipit population because they would have less food.

Removing a predator species, such as hen harriers, might lead to increases in prey species such as red grouse and meadow pipits.

Hints & tips

Questions in this area often ask you to predict or suggest the effect that the removal of one species from the food web will have on another species and may ask you to justify your prediction. Possible answers include:

- **increase** due to removal of a predator or due to increased food availability or decreased competition
- **decrease** due to decreased food availability or increased competition for food or predation
- **stay the same** due to a combination of the above.

Key words

Biodiversity – the variety and relative abundance of species present in an ecosystem
Carnivore – consumer which eats other animals only
Community – all the organisms living in a habitat
Competition – interaction between organisms seeking the same limited resources
Consumer – animal which obtains energy by eating other organisms
Ecosystem – natural biological unit composed of habitats, populations and communities
Habitat – the place where an organism lives
Herbivore – consumer which eats plants only
Interspecific competition – competition between organisms of two different species for a common resource
Intraspecific competition – competition between organisms within the same species
Niche – the role an organism has within its community in an ecosystem
Omnivore – consumer which eats both plants and animals
Population – the number of organisms of the same species living in a defined area
Predation – obtaining food by hunting and killing prey organisms
Predator – animal which obtains food by hunting and killing prey organisms
Prey – animal which is hunted and eaten by a predator
Producer – green plant which makes food by photosynthesis
Species – a group of organisms with similar characteristics which can interbreed and produce fertile offspring

Questions ?

A Short-answer questions

1 A freshwater loch is an example of an ecosystem.
   Describe what is meant by the term 'ecosystem'. (2)
2 State the term which describes the role that an organism plays within its community. (1)
3 Describe what is meant by the term 'biodiversity'. (1)
4 Give the meanings of the following terms:
   a) habitat
   b) community. (2)
5 Draw a food chain diagram for grass plants that are grazed by rabbits which are preyed upon
   by foxes. (1)
6 The diagram shows a simple food chain for an oak wood:
   oak leaf → caterpillar → blue tit → sparrowhawk
   Name the following from the food chain:
   a) a producer (1)
   b) a primary consumer (1)
   c) a carnivore. (1)
7 Describe what is meant by the following terms:
   a) competition
   b) predation. (2)

B Longer-answer questions

1 Describe the differences between interspecific and intraspecific competition. (2)
2 Give an account of the factors that define an organism's niche. (3)
3 Predict what would happen to the populations of grouse **and** golden eagles if hen harriers were
   to be removed from the moorland ecosystem in Figure 3.2 (on page 87). Justify your answers. (4)

Answers on page 117

Key Area 3.2
Distribution of organisms

Key points

1 **Abiotic factors** are physical factors such as light intensity, temperature, pH and moisture. ☐

2 Abiotic factors can be measured using instruments such as light meters and moisture meters designed to give numerical values. ☐

3 **Biotic factors** are biological factors such as competition for resources, disease, food availability, **grazing** and predation. ☐

4 Biodiversity and the distribution of organisms in an ecosystem are affected by abiotic and biotic factors. ☐

5 A **sample** can be taken to represent a population of an organism in a habitat or an ecosystem. ☐

6 Large numbers of samples must be taken so that they are **representative** and **reliable**. ☐

7 A **transect** is a line across a habitat along which samples and measurements are taken. ☐

8 **Quadrats** can be used to sample organisms that are fixed, or partly fixed, to a surface or those that spend a significant time motionless. ☐

9 **Pitfall traps** can be used to sample invertebrate animals that move around on surfaces. ☐

10 Sampling techniques such as those involving quadrats or pitfall traps are limited in what they can measure and can potentially create **sources of error**. ☐

11 A **paired statement key** is a set of contrasting questions which can be answered to identify and name a living organism. ☐

12 The presence, abundance or absence of **indicator species** can reveal information about the environment and level of pollution present. ☐

Summary notes

Abiotic factors

Abiotic factors are non-living or physical factors that affect organisms, such as light, moisture, temperature, pH, wind speed and oxygen availability.

Measuring abiotic factors

An abiotic factor is a non-living factor often related to climate that can affect the distribution of organisms in an ecosystem. There are many modern instruments that can be used to measure these factors. Most have a probe to contact the environment and an easily read scale to show the result. The table below shows some abiotic factors and methods used to measure them.

Abiotic factor	Measuring instrument
Soil pH	Soil pH meter
Light intensity	Light meter
Temperature	Thermometer
Soil moisture level	Moisture meter
Oxygen concentration	Oxygen meter

Figure 3.3 shows the type of meter that is commonly available in garden centres. These are often used by gardeners to obtain information on abiotic factors related to growing plants.

Notice the probe, which has to be cleaned and dried between readings, and the value recorded on the scale, which is easily read. It is good practice to average a large number of readings to help minimise error and increase reliability.

Figure 3.3 An example of a meter used to measure an abiotic factor

Sources of error

As in sampling, errors can creep into the measurement of abiotic factors. The techniques need to be applied with care, for example remembering to clean and dry probes between samples and reading scales accurately. Appropriately large numbers of readings are required for reliable averages to be calculated.

 Measuring abiotic factors is a technique which you need to know for your exam.

Biotic factors

Biotic factors are related directly to living organisms and include grazing, competition for food and space, predation and disease.

Sampling organisms in an ecosystem

It is usually impossible to count all the plants and animals living in an ecosystem so biologists use various methods for estimating the types and numbers of organisms present.

A sample can be taken to represent a population of organisms in an ecosystem. To be representative and reliable, appropriate numbers of samples are required. Normally, sample numbers and sizes should be as large as is safe and practicable.

Techniques used for sampling organisms

Quadrats are used to sample low-growing plants and very slow-moving animals. A quadrat marks off an exact area of ground so that

Hints & tips ★

You should now try Q4 in the Skills of Scientific Inquiry section on page 135 which covers this technique.

Hints & tips ★

Remember: ROAR = Repeat, Obtain an Average to increase Reliability

the organisms in that area can be identified and counted (Figure 3.4). Quadrats are often placed randomly in an area to improve reliability.

Pitfall traps can be used to sample small invertebrate animals living on the soil surface or in leaf litter (Figure 3.5). Animals fall into the trap and are unable to climb out again. The traps are usually set in a random way. It is essential to check traps regularly since predation and scavenging can occur within traps if left too long.

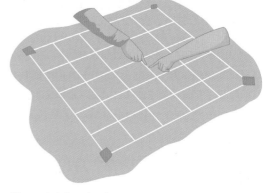

Figure 3.4 Quadrat in use

stones supporting cover to prevent rain flooding the trap and birds or other predators from removing the trapped animals

jar or pot sunk in a hole and with its lip level to the surface of the ground

Figure 3.5 Pitfall trap in use

Quadrats and pitfall traps can be used to sample the distribution of organisms in their habitats.

Distribution is often measured in relation to a changing factor such as light intensity or moisture. A study could be undertaken to investigate the effect of light intensity on the distribution of different species of plants between a field and the edge of a woodland or the effect of exposure on the distribution of seaweeds between high-tide and low-tide marks on a rocky shore.

A **transect** is a line of study across a habitat or part of a habitat, often marked by a long measuring tape. Figure 3.6 shows a transect line passing from a grassland area into a woodland. Quadrats can be placed at regularly spaced sampling stations along the line, and the number of each species of organism present in the quadrat identified and counted.

T Using a **transect line** is a technique which you need to know for your exam.

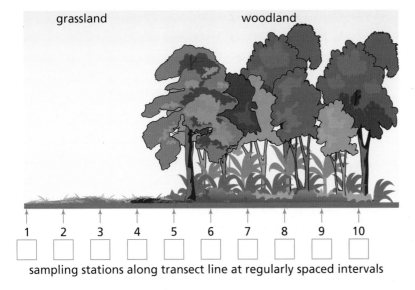

grassland   woodland

1   2   3   4   5   6   7   8   9   10

sampling stations along transect line at regularly spaced intervals

Figure 3.6 A transect line

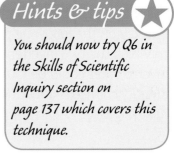

Hints & tips ⭐

You should now try Q6 in the Skills of Scientific Inquiry section on page 137 which covers this technique.

93

Possible limitations and sources of error in using sampling techniques

It is important that investigators plan their sampling, choosing the best sampling technique and the optimum number of samples for their purposes. The main sources of error usually lie with poor selection of technique or numbers of samples taken and with human error in carrying out the technique.

In general, large numbers of random samples should be taken and extreme care used when carrying out the sampling and identifying organisms present in the samples. Results should always be treated carefully and their reliability taken into account. The table below shows some limitations and sources of error in using quadrat and pitfall trap sampling.

Technique	Limitations	Possible errors
Quadrat sampling	Generally only suitable for low-growing, rooted plants or for slow-moving or motionless animals Quadrat size Reliability limited by number of samples possible	Quadrats may not be placed randomly or according to the planned method Inappropriate size of quadrat selected Too few quadrats used Plants misidentified, overlooked or miscounted
Pitfall trap sampling	Generally only suitable for small, surface-crawling invertebrates Pitfall trap size Reliability limited by number of traps set	Traps may not be placed randomly or according to the planned method Inappropriate size of trap selected Too few traps used Traps badly set or not emptied on time; invertebrates misidentified or lost from sample

Hints & tips

Remember: Repeat, Random, Reliable and Representative

You need to be familiar with techniques for **measuring the distribution of species** for your exam.

Hints & tips

You should now try Q5 in the Skills of Scientific Inquiry section on page 136 which covers this technique.

Paired statement keys

Paired statement keys can be used to identify species found in samples. The key on the following page can be used to identify the species to which the two invertebrates shown in Figure 3.7 belong. These invertebrates could be found during sampling procedures. Can you identify animals X and Y? (Check your answer at the bottom of the next page.)

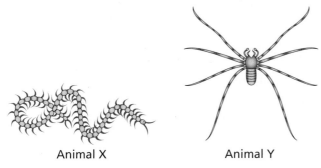

Animal X   Animal Y

Figure 3.7 Two unidentified invertebrates

Paired statements

1 three pairs of legs .... go to 3

   more than three pairs of legs .... go to 2

2 four pairs of legs .... **arachnid**

   more than four pairs of legs .... go to 4

3 two pairs of large scaly wings .... **lepidopteran**

   one pair of wings .... **dipteran**

4 seven pairs of legs .... **isopod**

   more than seven pairs of legs .... go to 5

5 one pair of legs per segment .... **chilopod**

   two pairs of legs per segment .... **diplopod**

Biodiversity

Biodiversity is the term used to describe the variety and relative abundance of species present in an ecosystem.

Factors affecting the biodiversity in an ecosystem

Biodiversity is affected by the following factors:

- abiotic factors
- biotic factors
- human influences, which can be abiotic or biotic.

Abiotic factors are non-living or physical factors that affect organisms and include light, moisture, temperature, pH, wind speed and oxygen availability.

Biotic factors are related directly to living organisms and include factors such as grazing, competition for food and space, predation, parasitism and disease. Competition for these resources can be intraspecific or interspecific. Intraspecific competition occurs between members of the same species. Interspecific competition occurs between members of different species.

Answer to question about the animals in Figure 3.7: X – chilopod; Y – arachnid.

Human activities affect biodiversity because they influence both abiotic factors, such as pH and temperature, and biotic factors, including grazing and predation. The following table summarises some human activities and their effects on biodiversity.

Human activity	Example	Effect on biodiversity
Habitat destruction	Deforestation of tropical rainforest	Total removal of biodiversity from an area, leading to climate change
	Overgrazing of natural dry grassland by cattle	Loss of biodiversity, leading to soil erosion and desertification
Pollution	Burning of fossil fuels, polluting the atmosphere	Loss of lichens and other plants, which then affects food chains
	Discharging untreated sewage into waterways	Deoxygenation of water and collapse of food chains
Hunting and fishing	Overhunting of game animals	Extinction of key species and wider effects on many food chains
	Overfishing for white fish	Loss of fish stocks and collapse of marine food chains
Conservation	Setting up wildlife reserves	Protects habitats and wild populations, restoring biodiversity
	Captive breeding of endangered species	Increases wild populations through release schemes

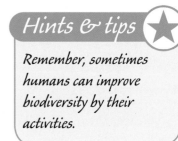

Hints & tips

Remember, sometimes humans can improve biodiversity by their activities.

Biological indicators of pollution

Organisms that, by their presence, abundance or absence, show conditions in the environment are called indicator species. Some can indicate the level of pollution present and are called biological indicators of pollution.

Freshwater invertebrates: water pollution

Certain invertebrates can be used to reveal information about the oxygen concentration in the freshwater environment. Rivers with water that is clean and well oxygenated will usually have a greater variety of different species than those with low oxygen concentration.

Mayfly nymphs and stonefly nymphs are examples of freshwater invertebrates whose presence indicates clean, well-oxygenated water conditions.

Rivers that have a low oxygen concentration have a small number of different species. Those organisms that can tolerate the low oxygen concentrations will increase in number. They have adaptations that allow them to survive in these low-oxygen conditions. Sludge worms contain haemoglobin, which helps them to absorb the limited oxygen in the water. Where these species are abundant, a high level of pollution is indicated. Figure 3.8 shows some freshwater invertebrates and the conditions they indicate.

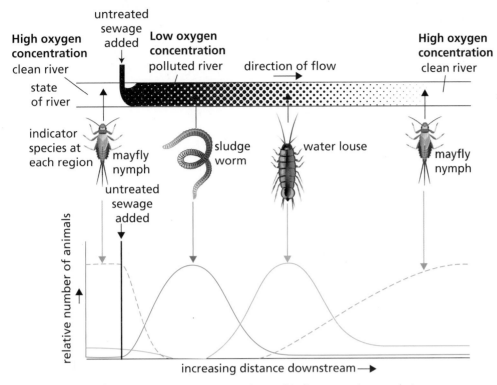

Figure 3.8 Effect of sewage on oxygen concentration and indicator species populations in a river

Lichen: air pollution

Lichen is an organism that is used as an indicator species for air pollution. Different species of lichen vary in their sensitivity to sulfur dioxide (SO_2) pollution produced by the burning of fossil fuels. Figure 3.9 shows the effect of sulfur dioxide pollution on the diversity of lichen species present in an ecosystem.

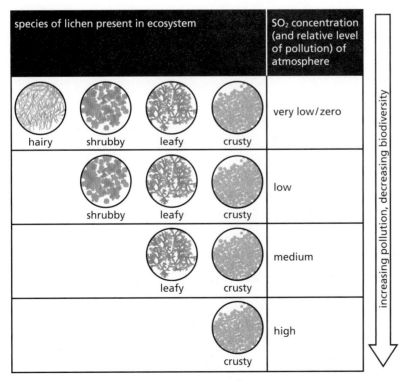

Figure 3.9 Effect of SO_2 pollution on the diversity of lichen species present in ecosystems

Key words

Abiotic factor – a physical factor, such as temperature or light intensity, that affects ecosystems

Biotic factor – factor related to the biological aspects of an ecosystem such as predation or competition

Grazing – method of feeding by which a herbivore eats grasses and herbs

Indicator species – organisms that by their presence, abundance or absence give information, such as level of pollution in the environment

Paired statement key – used to identify and name an unknown living organism

Pitfall trap – sampling technique used to trap animals living on the soil surface or in leaf litter

Quadrat – square frame of known area used for sampling the abundance and distribution of slow or non-moving organisms

Reliable – applies to results which can be trusted or in which there is confidence

Representative sample – a sample which is large enough and selected carefully enough to be truly proportional to the whole

Sample – a representative part of a larger quantity

Source of error – the origin of a mistake in drawing conclusions from experiments

Transect – a line of study across a habitat or part of a habitat, often marked by a long measuring tape

Questions ?

A Short-answer questions

1 Give **two** examples of abiotic factors in an ecosystem. (2)
2 Describe the following biotic factors: (2)
  a) predation
  b) competition
3 State what is meant by a sample. (1)
4 Name techniques that could be used to sample the following organisms: (2)
  a) dandelions rooted in a field of rye grass
  b) beetles that move around the soil surface at night in grassy areas
5 Give **two** sources of error that can affect sampling using quadrats. (2)
6 Give **two** sources of error that can affect sampling using pitfall traps. (2)
7 Give the meaning of the term 'indicator species'. (1)
8 Give **one** example of an indicator species. (1)

B Longer-answer questions

1 Explain how a named abiotic factor can affect the distribution of a green plant. (2)
2 Explain how a named abiotic factor can affect the distribution of an animal. (2)
3 Give **one** example of human influences that might affect biodiversity in an ecosystem and describe its effect. (2)
4 Explain what is meant by the term 'representative' as applied to samples. (2)
5 Explain what is meant by the term 'reliable' as applied to samples. (3)
6 Describe how you might estimate the number of dandelion plants per square metre in a large grassy park with an area of 100 m². (4)
7 Describe how you might measure the diversity of ground beetle species living in a small lawn with an area of 10 m². **T** (2)
8 Describe how you would measure the soil moisture level in an area of woodland. **T** (4)
9 Describe how you would measure the distribution of a species of seaweed on a rocky shore. **T** (4)

Answers on page 118

Key Area 3.3
Photosynthesis

Key points

1 Photosynthesis is the chemical process by which green plants make their own food. ☐
2 Photosynthesis is a series of enzyme-controlled reactions in two stages. ☐
3 The first stage of photosynthesis is light-dependent (**light reactions**). ☐
4 In the light reactions, light energy from the Sun is trapped by **chlorophyll** in the chloroplasts and is converted into chemical energy which is used to generate ATP. ☐
5 In the light reactions, water molecules are split to produce hydrogen and oxygen. Excess oxygen diffuses from chloroplasts and out of the cell. ☐
6 The second stage of photosynthesis is the **carbon fixation stage**. ☐
7 In the carbon fixation stage, hydrogen from the light reactions is combined with carbon dioxide to produce **sugar**. ☐
8 In the carbon fixation stage, energy from ATP is used in the production of sugar. ☐
9 The chemical energy in sugar is available for respiration. ☐
10 Sugar not used in respiration can be converted to plant **carbohydrates** such as **starch** or cellulose. ☐
11 Starch acts as a store of energy in plants while cellulose is a main structural component of plant cell walls. ☐
12 A **limiting factor** is a variable that, when in short supply, can limit the rate of a chemical reaction such as photosynthesis. ☐
13 The rate of photosynthesis is limited by various factors such as temperature, light intensity and the concentration of carbon dioxide available. ☐

Summary notes
Equation for photosynthesis

Photosynthesis is the process by which green plants make their own food. Carbon dioxide from the air and water from the soil are combined in green cells in the presence of light energy to produce sugar (glucose) and oxygen. This can be summarised by the following equation:

$$\text{water} + \text{carbon dioxide} \xrightarrow[\text{chlorophyll}]{\text{light energy}} \text{sugar (glucose)} + \text{oxygen}$$

Photosynthesis: a two-stage process

Photosynthesis is an example of an enzyme-controlled process in cells. It is a two-stage process.

First stage: light reactions

The first stage of photosynthesis is light dependent. Light energy from the Sun is trapped by pigment molecules such as chlorophyll in the chloroplasts of green plants. The main photosynthetic tissue in green plants is palisade mesophyll in leaves. The energy is then converted to chemical energy stored in molecules of adenosine triphosphate (ATP). The ATP is then passed to the second stage.

In the same set of reactions, water molecules are split into hydrogen and oxygen molecules. Excess oxygen diffuses from cells and is eventually released from the plant into the atmosphere.

Second stage: carbon fixation

The second stage of photosynthesis is completely temperature dependent because it is a series of enzyme-controlled reactions. ATP and hydrogen from the first stage are used with carbon dioxide that has been taken from the environment to produce sugar (glucose), which contains the chemical energy. This chemical energy is available for respiration, or the sugar can be converted to plant carbohydrates such as starch and cellulose.

The stages of photosynthesis are shown in Figure 3.10.

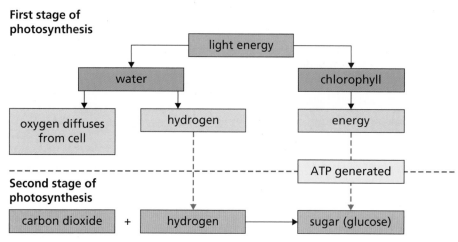

Figure 3.10 Two stages of photosynthesis

Investigating photosynthesis
Conditions for photosynthesis

A green plant leaf kept in the dark will become de-starched because it uses its starch up to provide it with energy when not photosynthesising. Leaves can be tested for starch using **iodine solution**, which turns from brown to blue-black with starch (Figure 3.11).

Step 1: Boil leaf in water then in alcohol

Step 2: Place leaf on tile then flood with iodine solution

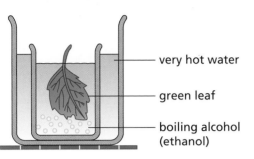

very hot water

green leaf

boiling alcohol (ethanol)

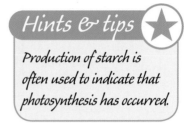

iodine solution – turns from brown to blue-black with starch

Figure 3.11 Testing a leaf for starch

A de-starched plant can be used to confirm factors needed for photosynthesis (Figure 3.12).

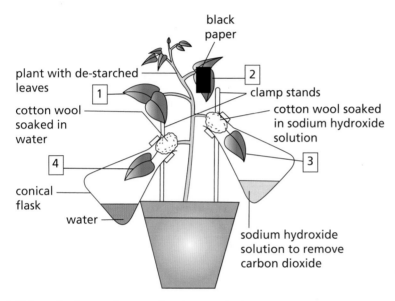

black paper

plant with de-starched leaves

1

2

clamp stands

cotton wool soaked in water

cotton wool soaked in sodium hydroxide solution

4

3

conical flask

water

sodium hydroxide solution to remove carbon dioxide

Figure 3.12 Investigating requirements for photosynthesis

Presence of starch in the leaves indicates that photosynthesis has occurred. The table below shows results of starch tests on the experimental leaves in Figure 3.12.

Leaf	Conditions given	Iodine test	Conclusion
1	Light and CO_2	Starch present	Photosynthesis has occurred
2	No light, CO_2	Starch absent	Photosynthesis requires light
3	Light, no CO_2	Starch absent	Photosynthesis requires CO_2
4	Light and CO_2 (control for leaf 3)	Starch present	Presence of flask for leaf 3 does not affect photosynthesis

Rate of photosynthesis

The rate of photosynthesis can be investigated using stems of aquatic plants such as *Cabomba*. Figure 3.13 shows an experimental set-up used in this type of investigation. Oxygen is a product of photosynthesis so its rate of production can be used as a measure of the rate of photosynthesis.

T **Measuring the rate of photosynthesis** is a technique you need to know for your exam.

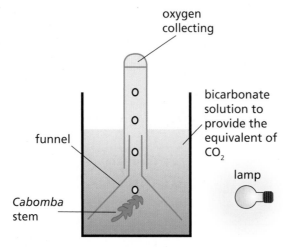

Figure 3.13 *Cabomba* bubbler

Limiting factors

The rate of chemical reactions such as photosynthesis can be limited by factors that are at low levels or are in short supply. Low temperature or light intensity can limit the rate of photosynthesis. The concentration of carbon dioxide available to the plant can also be limiting.

The graph in Figure 3.14 shows the effects of various combinations of temperature and carbon dioxide concentration on the rate of photosynthesis. Note how the graphs are very similar in shape.

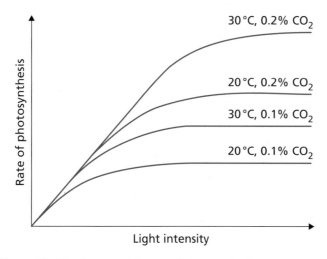

Figure 3.14 Effects of limiting factors on the rate of photosynthesis

Hints & tips
You should now try Q7 in the Skills of Scientific Inquiry section on page 138 which covers this technique.

Hints & tips
Measuring the rate of photosynthesis often involves measuring the rate of product production. It can also be measured by the rate of carbon dioxide usage by a plant.

Hints & tips
Measuring the **rate** of photosynthesis always requires a measurement of **time** to be made — bubbles of oxygen **per minute**.

Hints & tips
It is better to write about light **energy** or light **intensity** rather than simply using the word light.

Hints & tips
On limiting factor graphs, watch for the slope of the line — if the line slopes, then the factor on the x-axis is limiting the rate; where the line is level, some other factor is limiting the rate.

Key words

Carbohydrate – a substance such as sugar, starch, cellulose or glycogen, containing the elements carbon, hydrogen and oxygen

Carbon fixation stage – second stage in photosynthesis in which ATP, hydrogen and carbon dioxide are involved in the production of sugar

Chlorophyll – green pigment in chloroplasts that absorbs light energy for the process of photosynthesis

Iodine solution – brown-coloured solution that turns blue-black with starch

Light reactions – first stage in photosynthesis producing hydrogen and ATP required in the carbon fixation stage

Limiting factor – a variable that, when in short supply, can limit the rate of a chemical reaction such as photosynthesis

Starch – storage carbohydrate in plants

Sugar – energy-rich sub-units made by green plants in photosynthesis that can join into larger carbohydrates such as starch

Questions ?

A Short-answer questions

1   Name the two stages of photosynthesis.   (2)
2   Name the pigment in chloroplasts that traps the light energy required for photosynthesis.   (1)
3   In the light reactions of photosynthesis light energy is converted to chemical energy. Name the molecule in which the chemical energy is stored.   (1)
4   In the light reactions some energy is used to split water molecules. State what happens to the oxygen and hydrogen produced by this split.   (2)
5   State the role of carbon dioxide in the carbon fixation stage of photosynthesis.   (1)
6   Give **two** possible limiting factors in the process of photosynthesis.   (2)

B Longer-answer questions

1   Name the raw materials and requirements for photosynthesis and give the products of this process.   (5)
2   Describe **three** uses of the sugar produced by photosynthesis.   (3)
3   Explain what is meant by a limiting factor.   (2)
4   Explain why an increase in temperature can lead to an increase in the rate of photosynthesis.   (2)
5   Describe the need for the control leaf 4 in Figure 3.12 (on page 101).   (2)
6   Give an account of the carbon fixation stage of photosynthesis in a green leaf cell.   (3)
7   Using the experimental set-up shown in Figure 3.13 (on page 102), describe the measurements required to measure the rate of photosynthesis in a stem of pondweed. **T**   (2)

Answers on page 119

Energy in ecosystems

Key points !

1. As energy passes from one level to the next in a **food chain**, the majority is lost as heat, movement or undigested materials. ☐
2. Only a very small quantity of energy is used for growth and is therefore available at the next level in a food chain. ☐
3. A **pyramid of numbers** is a diagram that represents the total numbers of organisms at each stage of a food chain. ☐
4. A pyramid of numbers diagram can be an irregular shape if it involves producers such as trees which have a large mass or parasites which are of individually small masses. ☐
5. A **pyramid of energy** is a diagram that shows the total energy in the bodies of all organisms at each level of a food chain. ☐

Summary notes

Energy flow in an ecosystem

The Sun is the ultimate source of energy for an ecosystem. Light energy from the Sun is trapped by green plants and converted to the chemical energy that maintains all life on Earth.

Green plants are called producers because they can produce their own food by the process of photosynthesis. Consumers obtain energy by eating other organisms. Those which eat plants only are herbivores. Those which eat animals only are carnivores. Omnivores eat both animals and plants. The flow or transfer of energy through different feeding levels in an ecosystem can be shown by the use of food chain diagrams.

Energy transfer in food chains

The energy in the food that is eaten by a consumer is used in a number of ways.

Growth

Some energy is used to build new body tissue. This energy is trapped in body tissues and is available to be passed on to the next level in the chain.

Heat and movement

A lot of the energy is used to maintain body temperature and more is used in moving around. All of this can be lost as heat. This energy is not available to the next level in the chain.

Undigested material and uneaten remains

Undigested waste material and uneaten remains contain chemical energy. This energy is lost from the food chain. Most of the energy that enters a food chain level is ultimately lost and only a very small quantity is used for growth and is available at the next level in the food chain. This general pattern is shown in Figure 3.15.

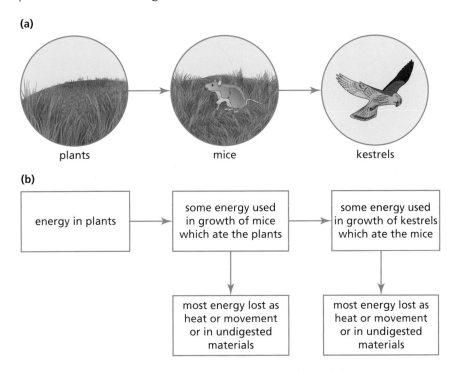

Figure 3.15 (a) A food chain (b) Energy flow and loss in the food chain

Pyramid diagrams

Pyramids of numbers

A pyramid of numbers is a diagram that shows the total number of organisms at each stage in a food chain (Figure 3.16).

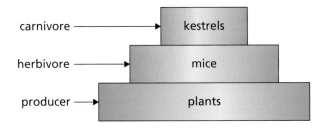

Figure 3.16 Typical regular pyramid of numbers

Moving up the pyramid, the number of organisms decreases but the size of each individual usually increases. The energy lost at each stage in the food chain limits the numbers of organisms that can survive at the next level. Pyramids of numbers do not always result in a true pyramid shape and can be irregular in shape. Figure 3.17 shows two examples of this.

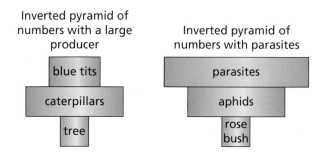

Figure 3.17 Two irregular pyramids of numbers

Pyramids of energy

A pyramid of energy is a diagram that shows the energy available at each stage in a food chain (Figure 3.18).

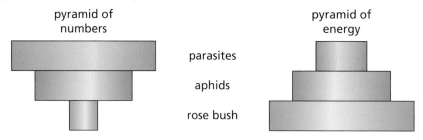

Figure 3.18 Irregular pyramid of numbers with its regular pyramid of energy

A pyramid of energy is the most realistic way of representing energy levels in a food chain and always has the shape of a true pyramid. Even food chains which have irregular pyramids of numbers can be represented as true pyramids of energy.

Key words

Food chain – simple set of feeding links showing how energy passes from a producer to consumers
Pyramid of energy – diagram that shows the relative quantities of energy at each level in a food chain
Pyramid of numbers – diagram that shows the relative numbers of organisms at each level in a food chain

Questions ?

A Short-answer questions

1 Pyramid diagrams are used to represent food chains. Explain what is represented by the following pyramids:
    a) numbers
    b) energy   (2)

B Longer-answer questions

1 Account for the loss of energy at each level in a food chain.   (2)
2 Give an example of a food chain of three species which would be represented by an irregular pyramid of numbers but a regular pyramid of energy.   (3)

Answers on page 119

Key Area 3.5

Food production

Key points !

1 The increasing human population requires an increased food supply. ☐

2 Methods of increasing food yield have included the use of **fertilisers**, **pesticides**, **biological control** and genetically modified (GM) crops. ☐

3 Fertilisers can be added to the soil to provide chemicals such as **nitrates** which increase crop yield. ☐

4 Nitrates dissolved in soil water are absorbed into plants. ☐

5 Nitrates are used to produce amino acids which are synthesised into plant proteins. ☐

6 Animals consume plants or other animals to obtain amino acids for protein synthesis. ☐

7 Fertilisers can leach from fields into freshwater, adding extra, unwanted nitrates, increasing the effects of **algal blooms**. ☐

8 These increased algal blooms reduce light levels, killing aquatic plants. ☐

9 These dead plants and dead algae are used as a food supply and are decomposed by **aerobic** bacteria, which multiply and reduce the oxygen concentration in the water. ☐

10 The decrease in oxygen concentration results in the death of many other organisms and reduces biodiversity. ☐

11 GM crops can be used to reduce the use of fertilisers. ☐

12 Pesticides are chemicals that can be used to kill plants and animals which reduce crop yield. ☐

13 Pesticides sprayed onto crops can accumulate in the bodies of organisms over time and so **bioaccumulate** along food chains. ☐

14 **Toxicity** levels of pesticides in the bodies of carnivores can build up and reach lethal levels. ☐

15 Biological control relies on natural solutions to pest problems, such as making use of natural predators or parasites. ☐

16 Crop species can be genetically modified to increase their yields without the need for fertilisers or pesticides. ☐

17 Biological control and GM crops are alternatives to the use of pesticides. ☐

18 Using biological control and GM crops rather than pesticides could result in less damage to the environment. ☐

Summary notes

Human population

The human population is continuing to increase, as shown in Figure 3.19. The milestones show the dates at which the population had added a further billion. Some cultural changes that contributed to population rise are also shown.

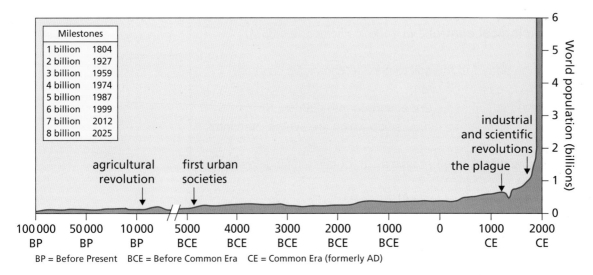

Figure 3.19 Human population growth, with milestones

To provide enough food to meet the needs of a rising population, methods of increasing food yield are needed. Farming is the main way in which humans guarantee food, and much effort has gone into increasing yields from farms.

Intensive farming

Intensive farming usually involves growing single crop species, such as wheat, in enormous fields. This allows efficient planting, crop treatments and harvesting. It is associated with heavy use of fertilisers and pesticides.

Fertilisers and pesticides

Fertilisers are often used to provide chemicals such as nitrates in the soil and improve the growth of the crop, so increasing yield. Various pesticides can be applied to the crop to kill a range of pest species (plants and animals) and so increase the amount of crop available to be harvested.

Problems in the use of fertilisers in freshwater ecosystems

Fertilisers can leach from the fields into the freshwater of rivers, lochs and ponds. This can happen when heavy rainfall washes fertiliser from fields into water as shown in Figure 3.20. The fertiliser greatly increases the growth of algae in the water and increases the effects of algal blooms on the freshwater ecosystem. An algal bloom is an abundance of algal cells, which can block out light to lower levels of a body of water and cause the death of other water plants.

In autumn, nutrients run out and there is less light for photosynthesis. The algae in the bloom start to die and their bodies and the bodies of other dead water plants are decomposed by aerobic bacteria. The bacteria increase in number and so reduce the oxygen concentration in the water. The decrease in oxygen concentration results in the death of many other organisms and results in a reduction in biodiversity.

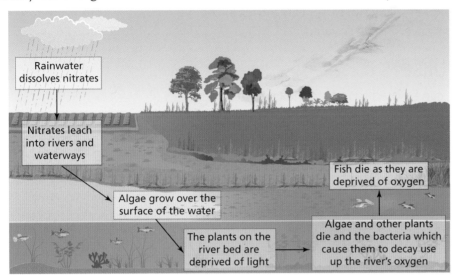

Figure 3.20 The effects of fertilisers in freshwater ecosystems

Problems in the use of pesticides

Pesticides are chemicals that are sprayed onto crops to kill pests. Herbicides are used to kill weeds and fungicides are used to kill fungal infections. Insecticides kill insects, such as greenfly, that feed on plants and reduce crop yield. Some insecticides are not specific and kill other insects as well as the target pest. This then affects other organisms that feed on the insects and therefore affects other food chains. Some pesticides are persistent and can build up, accumulating in the bodies of organisms higher up the food chain. This is called bioaccumulation. The toxicity levels of pesticide in the bodies of carnivores can build up to concentrations that can eventually reach lethal levels or damage their reproduction, as shown in Figure 3.21.

Hints & tips

Fertilisers Feed crops

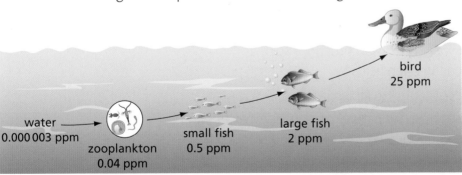

Figure 3.21 Bioaccumulation of the insecticide DDT along a food chain

The story of DDT

One of the most well-known insecticides is DDT, which was used very successfully from the 1940s and was responsible for controlling populations of some species of mosquito in certain areas of southern Europe, where these insects were involved in the spread of malaria.

This undoubtedly saved millions of human lives. However, DDT is non-selective and persistent, so it killed insects other than mosquitoes and accumulated in the food chain. Every molecule of DDT ever used is still somewhere in the world's ecosystems! One consequence of the use of this insecticide was very severe damage to populations of predatory birds and some species were brought near to extinction. As new insecticides became available, DDT was banned in many developed countries in the 1970s. Predatory bird populations have since recovered.

Alternatives to fertilisers

Genetically modified (GM) crops can be used to reduce the use of fertilisers.

'Genetically modified' or 'genetically engineered' are terms which refer to organisms whose genetic information has been altered, usually by the addition of a useful gene from another organism. Common GM foods include tomatoes, maize, rice, cabbage, potatoes and soya beans.

Crop plants can be genetically modified to increase their yields and also to reduce the need to use fertilisers.

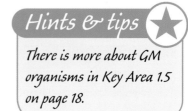

Hints & tips

There is more about GM organisms in Key Area 1.5 on page 18.

Alternatives to pesticides

Biological control methods and the use of GM crops are alternatives to the use of large amounts of pesticides to increase crop yield. Using biological control and GM crops could reduce the use of these pesticides and result in less damage to the environment.

Genetically modified herbicide-resistant crops can be sprayed with herbicide to kill weeds and so reduce competition for water and minerals. Other examples of GM crops include a variety of cauliflower that has been genetically modified by the insertion of scorpion genes. These produce a poison that acts as an insecticide and kills the caterpillars that are pests of the cauliflower crop.

Biological control

Biological control may be an alternative to the use of pesticides and involves using organisms such as natural predators or parasites to control pest numbers. This can work well in enclosed situations such as glasshouses but is more difficult in open fields.

Biological pest control works particularly well when the pest has been introduced to the ecosystem and has no natural predators. An example is the cotton cushion scale insect, which was accidentally introduced to California from Australia in the late nineteenth century. In California it multiplied out of control and destroyed large numbers of citrus trees, a major Californian crop. So a ladybird beetle, one of the scale insect's natural predators, was also introduced from Australia, and quickly reduced the numbers of scale insects to a safe level. Today both species coexist in California, but at low population densities.

The control species have to be chosen carefully to ensure that they can survive in their new environment and attack the pest only, not other native species. It is vital that they do not carry disease into the new ecosystem and that they will not become a pest due to lack of predators or parasites controlling their numbers.

Control species should be trialled in a controlled area, such as a greenhouse, before being released into the wild. If proper precautions are not taken, biological control can lead to ecological disaster.

Ecological problems caused by biological control

Cane toads were introduced to Australia from Hawaii in 1935 to control beetles that feed on sugar cane crops. The cane toads were poisonous to predators and ate a variety of prey, including native marsupials. They are still spreading through Australia and are now more of a problem to biodiversity than the original beetles.

Guppies are fish which eat mosquito larvae. They have been introduced to many parts of the world in the hope that they might control mosquitoes by eating their larvae, and so reduce the incidence of malaria. Although this has worked to some extent, the guppies usually have a negative effect on the native fish species by out-competing them.

Integrated pest management (IPM)

Some farmers use integrated pest management, which involves using a combination of biological control and pesticides to control crop pests. This method can help to reduce the amount of pesticide that farmers need to use.

Key words

Aerobic – in the presence of, or involving the use of, oxygen
Algal bloom – rapid increase of algal numbers in water
Bioaccumulation – the build-up of toxic substances in living organisms
Biological control – natural control of pests using natural predators, parasites etc.
Fertiliser – chemical added to the soil to improve plant growth or crop yield
Nitrates – nutrients absorbed from soil by plants to produce amino acids
Pesticides – general name for chemicals used to kill organisms that damage or feed on crop plants
Toxicity – poison level

Questions ?

A Short-answer questions

1 State how each of the following improves crop yield in intensive farming methods:
   a) fertilisers (1)
   b) pesticides. (1)
2 State why nitrates are needed by living organisms. (2)
3 Give the meaning of the term 'bioaccumulation'. (1)
4 State what is meant by the term 'algal bloom'. (1)
5 Name the gas that decreases in the water when aerobic bacteria decompose dead algae. (1)

B Longer-answer questions

1 Describe the link between farming methods and damage to the environment caused by the use of fertilisers. (5)
2 Describe the link between farming methods and damage to the environment caused by the use of persistent pesticides. (3)
3 Describe the advantages of using GM crops as a method of increasing food supply. (2)
4 Describe the advantages of using biological control of pests to increase food supply. (2)

Answers on page 120

Key Area 3.6
Evolution of species

Key points (!)

1 A **species** is a group of organisms that freely interbreed to produce fertile offspring. ☐
2 **Mutations** are random, spontaneous changes in genetic material and the only source of new alleles. ☐
3 Rates of mutation can be increased by environmental factors such as certain types of **radiation** and some chemicals. ☐
4 Mutations can be neutral and have little effect on the organism with the mutation. ☐
5 Mutations can be harmful if they give the organism a disadvantage and so decrease its chances of survival. ☐
6 Mutations can be beneficial if they give the organism an advantage and so increase its chances of survival. ☐
7 New alleles produced by mutation can result in plants and animals becoming better adapted to their environment. ☐
8 Variation within a population of living organisms makes it possible for a population to evolve over time in response to changing environmental conditions. ☐
9 Species often produce more offspring than the environment can support. ☐
10 The intensity of a selective factor such as predation is called **selection pressure**. ☐
11 **Natural selection** or survival of the fittest occurs when there are selection pressures. ☐
12 An **adaptation** is an inherited feature of an organism that makes it well suited to survive in its habitat. ☐
13 The best-adapted individuals in a population (those with the most favourable characteristics) have a **selective advantage**. ☐
14 Individuals with a selective advantage survive to reproduce, passing on the favourable alleles that confer the selective advantage to their offspring. ☐
15 The favourable alleles increase in frequency within the population. ☐
16 **Speciation** is the evolution of two or more species from one original ancestor species. ☐
17 Speciation occurs after part of a population becomes isolated by an **isolation barrier**, which can be geographical, ecological or behavioural. ☐
18 Different mutations occur in each sub-population. ☐
19 Natural selection selects for different mutations in each sub-population, due to different selection pressures. ☐
20 Each sub-population evolves until they become so genetically different that they are two different species and can no longer interbreed to produce fertile offspring. ☐

Summary notes
Evolution

In the nineteenth century, Charles Darwin used the idea of natural selection to explain the process of evolution that has resulted in the variety of living organisms we see on Earth today.

Species

A species is a group of organisms that are so similar to one another that they can freely interbreed and produce fertile offspring. All species that exist today have come about by the process of evolution.

Variation

As you saw in Key Area 2.4, members of a species are varied and much of their variation is passed onto their offspring. The only way in which new variation can appear is when errors in genetic material arise through a process called mutation.

Mutation

Mutations are spontaneous, random events that cause changes to the genetic information of an organism. They can affect single genes or whole chromosomes. Mutation is the only source of new alleles in a population. Mutations occur at a very low frequency, but environmental factors, such as radiation and some chemicals, can increase the rate of mutation.

Mutations can be neutral and have little effect on an organism. Mutations can be harmful and give the organism a disadvantage and so decrease its chance of survival. Alternatively, mutations can confer an advantage by giving an adaptation to the environment and increasing an organism's chances of survival.

Natural selection

Organisms produce more offspring than the environment can support due to the limited resources available. Members of a species show variation in their characteristics through the different mutations they carry. A struggle for survival follows as offspring compete for limited resources such as space, food and light.

Natural selection results in the survival of those organisms whose variation makes them better adapted to their environment. Individuals with favourable adaptations have a selective advantage, allowing them to survive – this is survival of the fittest. Selection pressure is a factor that acts on members of a population and results in the death of some members of the population and the survival of others. Examples of selection pressures include predation, disease and various aspects of the climate such as rainfall and temperature.

Speciation

Speciation is the term used to describe the formation of two or more species from one original species. The process happens in a series of stages:

1 **Isolation:** isolation barriers prevent breeding between sub-populations of a species.

   a) **Geographical barriers:** Barriers can be geographical as shown in Figure 3.22(a). The squirrels do not cross the canyon and so the two separated sub-populations do not interbreed.

   b) **Ecological barriers:** Barriers can be ecological as shown in Figure 3.22(b). Sub-populations of a species which are adapted to ecological differences, such as pH or salinity conditions, or to different habitats, are separated and do not meet to interbreed. The lizards in different habitats within the trees do not interbreed.

   c) **Behavioural barriers:** Barriers can be behavioural as shown in Figure 3.22(c). The sub-group of flies which prefer papaya cannot breed with the sub-group which prefer banana.

(a) A species of ground squirrel interbreeds freely across an open plains area.

A break in the Earth's crust opens up a canyon between two sub–populations. This forms a barrier to breeding between squirrels on either side of the canyon.

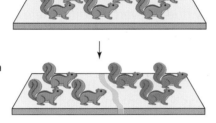

(b) A species of lizard interbreeds freely in the high branches of trees.

Some individuals change their habitat preference and move down to the lower areas of trees. This forms a barrier since the two sub-populations hardly ever meet.

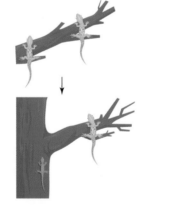

(c) A species of fruit fly interbreeds freely because all individuals mate on, and lay eggs in, banana fruit.

Some individuals change their preference and only mate on, and lay eggs in, papaya fruit. This forms a barrier to breeding between the papaya flies and the banana flies.

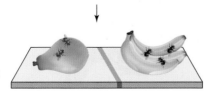

Figure 3.22 Examples of isolation barriers in **(a)** ground squirrels, **(b)** lizards and **(c)** fruit flies

2 **Mutation:** different mutations are likely to arise in the different sub-populations. Advantageous mutations occurring in one sub-population will allow the mutants to survive, reproduce and pass on their beneficial alleles to their offspring.

3 **Natural selection:** the environment of each sub-population is different, as are the selection pressures. Natural selection of the best-suited individuals in each sub-population is therefore different. Those with favourable characteristics have an advantage over others. They survive, reproduce and pass on these favourable characteristics to their offspring. In this way the sub-populations gradually evolve along different paths.

4 **Speciation:** after a very long period of time, the isolated sub-populations change so much genetically that they can no longer interbreed to produce fertile offspring even if they are able to reach each other. This shows that they are completely new species. Speciation has occurred.

Speciation in action: Darwin's finches

The Galapagos Islands are isolated in the Pacific Ocean 600 kilometres from the South American coast. It is thought that a species of finch-like bird arrived on these islands hundreds of thousands of years ago after being forced to leave the mainland, probably due to overcrowding there. The birds spread out over the islands and lack of competition allowed their populations to rise.

Groups became isolated on the individual islands because it was dangerous to cross the windy and choppy seas. Different mutations occurred within the populations of different islands and natural selection varied between the islands because of habitat differences. Over thousands of generations, differences between populations built up and today there are about 13 new species of Darwin's finches. The finch species differ strikingly in terms of beak shapes and sizes, allowing them to exploit particular foods available in their habitat, as shown in Figure 3.23.

insect-eating 'woodpecker' finch (tree dweller)

insect-eating 'warbler' finch (tree dweller)

fruit-eating finch (tree dweller)

seed-eating finch (ground dweller)

cactus-eating finch (ground dweller)

Figure 3.23 Darwin's finches

Key words

Adaptation – feature of an organism that helps it to survive

Isolation barrier – barrier that prevents interbreeding and gene exchange between sub-populations of a species

Mutation – a random and spontaneous change in the structure of a gene, chromosome or number of chromosomes; only source of new alleles

Natural selection – process which ensures the best adapted individuals survive

Radiation – energy in wave form such as light, sound, heat, X-rays, gamma rays

Selection pressure – factor such as predation or disease that affects a population, resulting in the death of some individuals and survival of others

Selective advantage – an increased chance of survival for an organism because of possession of favourable characteristics

Speciation – formation of two or more species from an original ancestral species

Species – organisms with similar characteristics and with the ability to interbreed to produce fertile offspring

Questions

A Short-answer questions

1 Give a definition of the term 'species'. (1)
2 State what is meant by genetic mutation. (1)
3 Name **two** factors that can increase the rate of mutation. (2)
4 State what is meant by an advantageous mutation. (1)
5 State what is meant by the term natural selection. (1)
6 Give the meaning of the term 'adaptation'. (1)

B Longer-answer questions

1 Explain the importance of mutations in the evolution of new species. (2)
2 Explain what is meant by the terms 'selection pressure' and 'selective advantage'. (2)
3 Give an account of speciation. (4)

Answers on page 120

Answers

Key Area 3.1

A Short-answer questions

1 habitat; and community   [1 each] (2)
2 niche (1)
3 variety/number/relative abundance of species (1)
4 a) place in which an organism lives (1)
   b) all of the living organisms in a habitat (1)
5 grass → rabbit → fox (1)
6 a) oak leaf (1)
   b) caterpillar (1)
   c) blue tit **OR** sparrowhawk (1)

7 a) competition occurs when organisms struggle between or within species for resources that are in short supply (1)
   b) predation is a relationship in which one animal hunts and eats another (1)

B Longer-answer questions

1 interspecific competition is between organisms of two different species for a common resource; intraspecific competition is between organisms within the same species   [1 each] (2)

⇨

2 what it feeds on; what feeds on it; what other resources it uses; what organisms it interacts with; what it adds to the environment   [any 3]   (3)

3 red grouse: increase; decreased predation   [1 each]
   golden eagle: increase; increased food availability **OR** decreased competition for food   [1 each]   (4)

Answers

Key Area 3.2

A Short-answer questions

1 temperature; pH; light intensity; moisture (other possible answers)   [any 2]   (2)

2 a) predation is a relationship in which one animal hunts, kills and eats another   (1)
   b) competition occurs when organisms struggle against others for similar limited resources   (1)

3 a small, representative portion of a larger quantity   (1)

4 a) quadrat   (1)
   b) pitfall trap   (1)

5 failure to use enough quadrats; using inappropriately sized quadrats; failure to randomise the quadrats   [any 2]   (2)

6 failure to use enough traps; inappropriately sized traps; failure to randomise distribution of traps; failure to check traps regularly; trap lip not flush with soil level   [any 2]   (2)

7 species whose presence, absence or abundance suggests information about the state of the environment   (1)

8 mayfly; lichen (other possible answers)   (1)

B Longer-answer questions

1 light intensity can affect photosynthesis; temperature can affect enzyme activity in plant cells; low water/moisture/humidity/rainfall can cause loss of turgidity in cells and cause wilting in whole plants (other possible answers) [1 for factor + 1 for explanation]   (2)

2 temperature can affect enzyme activity in animal cells; low water/moisture/humidity/rainfall can lead to dehydration in animals (other possible answers)   [1 for factor + 1 for explanation]   (2)

3 SO_2 pollution/burning fossil fuels which damages plants; overgrazing which damages plant communities/causes erosion of soil; deforestation which results in loss of habitat; overhunting/overfishing which can result in extinction of key species (other possible answers) [cause = 1, effect = 1]   (2)

4 a sample which is large enough and accurately taken; accurately reflects/is proportional to the whole   [1 each]   (2)

5 randomly taken samples; which can be trusted or in which there is confidence; because enough/a sufficient number have been taken   [1 each]   (3)

6 drop many $1\,m^2$ quadrats randomly across the area; count dandelion plants in each quadrat; find the average number across all the quadrats used; multiply the average number in each quadrat by 100   [1 each]   (4)

7 set a number of pitfall traps randomly across the lawn; identify beetle species in each trap and count their numbers [1 each]   (2)

8 use a soil moisture meter; take random readings; repeat; clean/wipe probe between readings; obtain an average reading   [any 4]   (4)

9 set up/use/lay out a transect line; use quadrats; at regular intervals along the transect line; identify/record presence of seaweed species in each quadrat [1 each]   (4)

Answers

Key Area 3.3

A Short-answer questions

1 light reactions/light-dependent stage; carbon fixation stage   [1 each]   (2)
2 chlorophyll   (1)
3 ATP   (1)
4 oxygen diffuses into atmosphere; hydrogen passes to carbon fixation stage   [1 each]   (2)
5 combines with hydrogen to form sugar/carbohydrate   (1)
6 temperature; light intensity; CO_2 concentration   [any 2]   (2)

B Longer-answer questions

1 **raw materials:** water; carbon dioxide [1 each]
   requirements: light **OR** chlorophyll [any 1]
   products: sugar/glucose; oxygen [1 each]   (5)
2 used in respiration; converted to starch; converted to cellulose; converted to lipid; converted to protein   [any 3]   (3)

3 factor close to its minimum value or in short supply; holds the rate of photosynthesis in check/limits the rate of photosynthesis   [1 each]   (2)
4 raised temperature increases rate of enzyme action; enzymes needed for photosynthesis   [1 each]   (2)
5 used as control to compare with leaf 3; shows that enclosing in flask does not affect photosynthesis   [1 each]   (2)
6 CO_2 absorbed from atmosphere; combines with H from light stage; requires ATP from light stage; makes sugar; energy in ATP locked into sugar molecules   [any 3]   (3)
7 count the number of bubbles (of oxygen) produced **OR** measure the volume of oxygen produced; per minute/in a given period of time [1 each]   (2)

Answers

Key Area 3.4

A Short-answer questions

1 a) total number of organisms/ populations of each organism at each level of a food chain   (1)
   b) total energy available at each level of a food chain   (1)

B Longer-answer questions

1 energy lost by movement; heat; uneaten remains/undigested materials   [any 2]   (2)
2 tree → caterpillar → blue tit (others possible) **OR** rose bush → aphids → parasites (others possible) [1 mark for each species correctly placed]   (3)

Answers

Key Area 3.5

A Short-answer questions

1. a) add nitrates, which allows plants to make more protein for growth (1)
   b) kill species that damage crop growth (1)
2. nitrates are used to produce amino acids; amino acids are then synthesised into proteins (2)
3. the accumulation/build-up of substances, such as pesticides or other chemicals, in an organism/when an organism absorbs a substance at a rate faster than that at which the substance is lost (1)
4. abnormally high growth of algae (1)
5. oxygen (1)

B Longer-answer questions

1. fertilisers can leach from field into water; this increases algal growth/causes algal bloom; light blocked out and plants/algae die; bacteria feed on dead plants/algae and increase in number; bacteria use up oxygen/deoxygenate water; other water organisms die/reduces biodiversity [any 5, 1 mark each] (5)
2. pesticides accumulate along food chains; build up in the bodies of carnivores; damage carnivores or their reproduction [1 each] (3)
3. have bigger yield; better disease/pest resistance; better cold/drought tolerance; don't require pesticides (other possible answers) [any 2] (2)
4. remove pests that reduce crop yield; have lower environmental impact than pesticide (other possible answers) [1 each] (2)

Answers

Key Area 3.6

A Short-answer questions

1. group of organisms interbreeding to produce fertile offspring (1)
2. random change in genetic material (1)
3. radiation; chemicals [1 each] (2)
4. confers an advantage to the organism/increases survival chance of the organism (1)
5. survival of the fittest **OR** best adapted organisms/those with favourable characteristics survive (1)
6. special feature of an organism that gives it the ability to survive (1)

B Longer-answer questions

1. mutations increase variation; natural selection acts on variation to change species over time [1 each] (2)
2. a factor (such as predation/disease) that affects a population resulting in the death of some individuals; an increased chance of survival because of a favourable characteristic [1 each] (2)
3. members of a species become separated by isolation barriers/examples; different mutations occur in the sub-populations; different selection pressures act on the sub-populations; natural selection/survival of the fittest occurs; after many generations new species are formed [any 4] (4)

Practice assessment: Life on Earth (45 marks)

Section 1 (10 marks)

1 The table below contains examples of factors affecting biodiversity in a Scottish mixed woodland ecosystem. Which line in the table below correctly identifies examples of biotic and abiotic factors?

	Biotic factors	Abiotic factors
A	Sparrowhawks are predators	Red deer graze low branches of deciduous trees
B	Soil has an acidic pH	Sparrowhawks are predators
C	Water is frozen for part of the winter	Soil has an acidic pH
D	Red deer graze low branches of deciduous trees	Water is frozen for part of the winter

2 Competition for water occurs between individual plants of a barley crop. Which line in the table below describes this competition?

	Type of competition	Relative level of intensity
A	Interspecific	High
B	Interspecific	Low
C	Intraspecific	High
D	Intraspecific	Low

3 Which line in the table below describes two characteristics of mutations?

	Characteristic 1	Characteristic 2
A	Random	Spontaneous
B	Non-random	Predictable
C	Random	Predictable
D	Non-random	Spontaneous

4 The diagram below shows a pyramid of energy for a food chain.

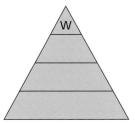

There is less energy at level W in the pyramid because:
A the mass of individual organisms at level W is low
B energy is stored in each level and is not passed on
C the energy at level W is concentrated in a small number of organisms
D energy is lost at each level in the food chain involved

5 Which method of increasing food production would reduce the occurrence of algal blooms and bioaccumulation?
A using pesticides and planting GM crops
B using fertilisers and pesticides
C using biological control methods and fertilisers
D planting GM crops and using biological control methods

6 What is the likely source of a selective advantage and the likely result for an organism?
   A mutation gives decreased chance of survival
   B mutation gives increased chance of survival
   C isolation barriers give decreased survival
   D isolation barriers give increased survival

7 The diagram below represents part of a food web in a freshwater ecosystem.

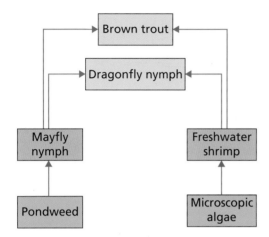

The population of mayfly nymphs in the ecosystem decreased dramatically following an unusually dry period.

Which prediction of the effect that this would have on pondweed and brown trout populations is supported by information in the diagram?
   A both populations would decrease
   B both populations would increase
   C pondweed population would decrease and brown trout population would increase
   D pondweed population would increase and brown trout population would decrease

8 The graph below shows the results of an investigation of the levels of insecticides in the muscle tissue of various farmland birds. The table shows information about the niches of these birds.

Land birds		Water birds	
Plant eating	Meat eating	Plant eating	Meat eating
Woodpigeon	Sparrowhawk	Moorhen	Grey heron

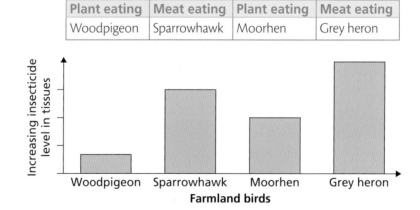

Which of the following conclusions is supported by the information in the graph?
   A insecticide damages water birds more than land birds
   B more insecticide is found in water habitats than in land habitats
   C meat-eaters accumulate more insecticide than plant-eaters
   D sparrowhawks eat woodpigeons

9 The information below refers to energy flow in a woodland ecosystem over a 24-hour period.
   ○ Light energy units available from the sun: 5 000 000
   ○ Energy units trapped by plants in photosynthesis: 10 000
   ○ Energy units released as heat during respiration: 9000

What percentage of available energy remained trapped in the ecosystem after this period?

A 0.02%

B 0.2%

C 10%

D 90%

10 The apparatus below was part of an experiment used to demonstrate the effects of SO_2 pollution on transpiration from a leafy shoot.

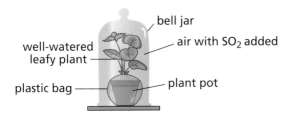

A suitable control for this experiment would be provided by using the same set-up but:

A using a black plastic bag

B leaving out the leafy shoot

C using SO_2-free air

D leaving out the bell jar.

Section 2 (35 marks)

1 The diagram below shows a food chain from a Scottish heather moorland ecosystem.

heather shoots → mountain hare → golden eagle

   a) Give **one** way in which energy can be lost at each level of this food chain. (1)

   b) Food chains like this one can be represented as pyramids of numbers. Describe what is meant by a pyramid of numbers. (1)

   c) The mountain hares are in competition for food with red grouse and with other mountain hares on heather moors. Copy and complete the table below to identify the type of competition in each case. (1)

Organism	Type of competition involved
Red grouse	
Other mountain hares	

2 Photosynthesis is a two-stage process by which green plants produce sugar.

   a) Name the first stage of photosynthesis. (1)

   b) Name **two** products of the first stage of photosynthesis which are used in the second stage. (2)

   c) Describe the second stage of photosynthesis. (3)

   d) Explain what is meant by a 'limiting factor' in photosynthesis. (1)

3 An investigation was carried out into the distribution of the alga *Pleurococcus* on the bark around the trunk of a mature tree. Six quadrats were placed around the circumference of the tree and the percentage of each occupied by the alga was measured. At the same stations on the tree, sample tubes were used to collect the moisture run-off from the bark in a 24-hour period. The results are shown in the table below.

Station	Moisture run-off from the bark (cm³ of water collected in a sample tube taped to the bark at each station per 24-hour period)	*Pleurococcus* (% cover)
1	0.5	24
2	0.8	32
3	1.2	54
4	1.8	80
5	1.9	85
6	2.1	98

a) On a piece of graph paper, plot a line graph to show the distribution of *Pleurococcus* against the moisture run-off from the bark. (2SSI)

b) Give the relationship between surface water run-off and the density of *Pleurococcus* on the bark of this tree. (1SSI)

c) Suggest how the reliability of the results for moisture run-off from the bark could be improved. (1SSI)

d) Calculate the percentage increase in the run-off between the lowest and highest values. (1SSI)

4 The following list refers to stages in an evolutionary process:

1 Natural selection

2 Mutation

3 Isolation

a) The diagram below shows the stages in the process and its end result. Copy the diagram and add numbers to the boxes to show the order in which the stages would normally occur. (1)

b) Give the name used for this type of evolutionary process. (1)

5 In recent decades human populations have grown and increases to food yield in agriculture have been needed. One method of increasing crop yields is by the use of fertilisers, which can leach into freshwater ecosystems and cause increased algal blooms.

a) (i) State what is meant by an algal bloom. (1)

(ii) Explain how algal blooms can lead to a reduction in oxygen in a freshwater ecosystem. (2)

b) The larval stages of insects such as mayfly and stonefly require high concentrations of oxygen in their environment. Their absence can be used to identify low concentrations of oxygen in a freshwater ecosystem.

(i) State the term used for organisms such as these. (1)

(ii) Describe a procedure using these organisms that can be used to gain information about concentrations of oxygen in a freshwater ecosystem. (2)

c) The table below shows the numbers of stonefly larvae found in four different months at a site on a freshwater canal.

Month	Number of stonefly larvae found in samples
February	23
May	25
August	12
November	47

⇨   (i) Describe **two** precautions that should be taken to ensure that the samples taken at
        each site can be validly compared.   (2SSI)

    (ii) Name the month during which the effects of the decay of algal bloom were most
        noticeable at the canal site and explain your choice.   (2SSI)

6 Pesticides have been used successfully to increase the yield of food crops but they have led to some
environmental damage.

Give **one** alternative to the use of pesticides which could be less damaging to the environment.   (1)

7 Give an account of what is meant by an organism's niche.   (2)

8 Read the following passage and answer the questions based on it.

Bioaccumulation in marine ecosystems

Toxic chemicals can enter marine ecosystems when industrial or agricultural waste runs off into rivers that
then flow into the sea. These pollutants can cause disease, genetic mutations, birth defects, reproductive
problems, behavioural changes and death in marine organisms.

The severity of the damage to species depends on their niche. In many cases, animals near the top of food
chains are most affected because of a process called bioaccumulation. Many of the most toxic chemicals
settle to the seabed and are taken in by organisms that feed on sediments there. These chemicals are not
degraded by enzymes. They accumulate within the animals that have taken them in and become more
and more concentrated as they pass along food chains as animals feed and then are eaten themselves.
This bioaccumulation means that predators build up greater and more dangerous concentrations of toxic
materials than animals lower in the food chain.

a) Give **two** examples of the effect of toxic chemicals on marine organisms.   (1)

b) Describe the niche of the marine animals which are most affected by the pollutants mentioned
in the passage.   (1)

c) Describe how the toxins enter marine food chains.   (1)

d) Explain why the toxins are able to bioaccumulate.   (1)

e) Suggest what could happen to the population of a prey organism if its predator population were
damaged more than the prey population by toxic chemicals.   (1)

Answers to practice assessment: Life on Earth

Section 1

1 D, 2 C, 3 A, 4 D, 5 D, 6 B, 7 D, 8 C, 9 A, 10 C

Section 2

1 a) heat **OR** movement **OR** undigested
    material   (1)

  b) a pyramid of numbers shows the total
    number/population of organisms at each
    level in a food chain   (1)

  c) interspecific
    intraspecific   [both]   (1)

2 a) light reactions   (1)

  b) ATP; hydrogen   [1 each]   (2)

  c) hydrogen is combined with carbon dioxide
    (to produce sugar); ATP supplies energy;
    (carbon fixation reactions) controlled by
    enzymes   [1 mark each]   (3)

  d) a factor which limits the rate at which
    photosynthesis occurs/sugar is
    produced   (1)

3 a) scales and labels; accurate plots and
    plots connected by straight
    lines   [1 each]   (2)

  b) as moisture run-off from the bark
    increases, the density/% cover of
    Pleurococcus increases   (1)

  c) repeating the experiment and
    calculating an average **OR** taking larger
    sample/using wider tubes   (1)

  d) 320%   (1)

4 a) 3 → 2 → 1   (1)

  b) speciation   (1)

⇨

⇨

5 a) (i) unusually high growth of algae in a
           body of water   (1)
      (ii) algae die; dead algae decomposed
           by aerobic bacteria, which use up
           oxygen   [1 each]   (2)
   b) (i) indicator species   (1)
      (ii) take samples in the river; count
           larvae; relate larvae to oxygen
           level   [any 2, 1 mark each]   (2)
   c) (i) same sampling method; same time
           spent on process; same weather
           conditions   [any 2]   (2)
      (ii) August; lowest stonefly count indicates
           lowest oxygen level   [1 each]   (2)
6 growing genetically modified/GM crops **OR**
   using biological control **OR** description
   of one of these   (1)

7 role played in habitat; what organism eats and
   what eats it; what resources it requires and
   what it adds to the habitat   [any 2]   (2)
8 a) disease/genetic mutations/birth defects/
      reproductive problems/behavioural
      changes/death   [any 2]   (1)
   b) predator   (1)
   c) settle to seabed and are taken in by
      organisms which feed on sediments   (1)
   d) because they are not degraded by
      enzymes   (1)
   e) population of prey could increase   (1)

Skills of Scientific Inquiry

The questions within the Key Area chapters of this book test demonstration of knowledge. This section covers the skills of scientific inquiry and includes questions to test these. We have given three different approaches to working with these science skills and recommend that you use all three.

- 4.1 gives an example of a scientific investigation and breaks it down into its component skills. There are questions on each skill area. The answers are on page 143.
- 4.2 goes through the skills one by one and gives you some exam-style questions covering the apparatus, techniques and skills you must be familiar with. There is a grid on page 130 showing which skills are tested in the parts of each question. The answers are given on pages 144–145.
- 4.3 provides sets of hints and tips on answering science skills exam questions.

4.1 An example

The *Cabomba* bubbler

Read through the information about the experiment below and then work through the skill areas listed, trying to comment on the questions in each category.

1 The apparatus was set up as shown in the diagram.

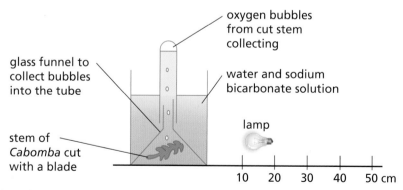

oxygen bubbles from cut stem collecting

glass funnel to collect bubbles into the tube

water and sodium bicarbonate solution

lamp

stem of *Cabomba* cut with a blade

10   20   30   40   50 cm

2 The plant was exposed to bright light and allowed to bubble for 10 minutes before the experiment was started.

3 The light intensity was altered by moving the lamp further away from the plant, and the rate of photosynthesis was measured by counting the number of bubbles of oxygen coming from the cut plant stem in a set period of time at each distance.

4 The experiment was repeated at each distance and the two results averaged.

5   The results obtained are shown in the table below.

Light intensity (distance of lamp from plant in cm)	Average rate of photosynthesis (bubbles of oxygen produced per minute)
10	20
20	20
30	15
40	10
50	5

Planning experiments and designing experiments

This is about confirming the aim of an experiment and suggesting a likely hypothesis, choosing apparatus, thinking about the dependent variable and deciding what to measure. Designing is closely related to planning but involves details of how often to measure, which variables need to be controlled and how to do this. It also involves anticipating possible errors and trouble-shooting these.

Q1   Suggest what the aim of the experiment was.

Q2   Suggest a hypothesis that would go with this aim.

Q3   Why is a water plant used?

Q4   Why is sodium bicarbonate solution used? (Hint: this solution acts as a source of carbon dioxide.)

Q5   Why are the bubbles counted? What's in the bubbles and why is this important?

Q6   Why should the plant be allowed to bubble for a while before the experiment starts? (Hint: *Cabomba* stems have natural air spaces, which help the plant float in water.)

Q7   How is light intensity altered?

Q8   Which vital piece of apparatus is missing from the diagram on page 127? (Hint: bubbles per *minute* are given in the results!)

Q9   Which variables should have been controlled?

Q10 Why was the experiment repeated at each light intensity?

Selecting information

This is about using a source such as a table or line graph to extract particular pieces of information. This can be simply reading off a value. The skill requires knowledge about labels, scales and units.

Q11 How many bubbles of oxygen are produced per minute when the lamp is 30 cm from the plant?

Q12 How far away from the plant was the lamp when the bubbling rate was 5 bubbles per minute?

Presenting information

This is about taking some information and presenting it in a different and more useful form. A common type of presentation might be to take information from a table and present it as a line graph. The skill requires knowledge of labels, scales and units as well as careful drawing using a ruler and accurate plotting.

Q13 Draw a line graph to show the rate of photosynthesis against the light intensity. Be careful to have the scale of light intensity going from least to most intense – think about it!

Processing information

This is often about working with numerical data and using calculations to convert a lot of data into a simple form.

Q14 What is the ratio of number of bubbles produced with the light at 10 cm to those produced at 50 cm?

Q15 What is the average rate of bubbling over all of the trials?

Q16 What is the percentage increase in the rate of bubbling when the lamp is moved from 50 cm to 20 cm from the plant?

Predicting and generalising

Predicting is about taking an experimental result and imagining what would happen if a variable changed. Generalising is about looking at experimental results and trying to find a rule that would hold true in all situations.

Q17 How would bubbling be affected by moving the lamp to 5 cm and 60 cm away?

Q18 How would you expect the results to differ if a larger size of water plant was used?

Q19 How would you expect the glucose concentration in the plant to change during the experiment?

Concluding and explaining

Concluding involves making a statement about the relationship between variables in an experiment. Explaining is about using knowledge to understand why a result has been obtained.

Q20 How is the rate of photosynthesis affected by light intensity?

Q21 Explain the results in terms of your knowledge about the light-dependent stage of photosynthesis and limiting factors in photosynthesis.

Evaluating

Evaluating is looking critically at an experiment and deciding if it is likely to be valid and if its results can be relied on; it is about looking for potential sources of error. Evaluating also involves suggesting improvements to an experiment that might remove sources of error in future experimental repeats.

Q22 What sources of error might be present?

Q23 Are all the bubbles the same size – does it matter? Is there a better way to measure the gas coming off?

Q24 How could the exact distance of the lamp from the plant be made certain? Is there a better way of changing light intensity?

Q25 Is any other light source present? Does that matter and how could you control this?

Q26 What factors, other than light intensity, might be limiting photosynthesis?

Q27 How reliable is one set of results with one plant and one lamp?

Q28 How could this experiment be improved? Make a list!

Q29 Which other factors that affect photosynthesis could be investigated using the *Cabomba* bubbler apparatus?

Q30 How could you adapt the experimental set-up to investigate one of these?

4.2 Skill by skill questions

The table below lists the basic skill areas that can be tested in your exam and the techniques you must be familiar with. We have provided seven practice questions (Q1–7) which cover all of the skills areas and all of the techniques, including most items of apparatus (in **bold** type within the questions), that you should be familiar with for your exam. The table shows the parts of the practice questions where you can find each skill tested. It is probably better to try the whole of each question in turn. If you find particular difficulty with any part of a question, you could use the table to identify the skill area that needs further work.

Skill area	Category within skill area	Techniques questions						
		Q1 Measuring enzyme activity	Q2 Using a respirometer	Q3 Using a potometer	Q4 Measuring abiotic factors	Q5 Measuring species distribution	Q6 Using a transect line	Q7 Measuring rate of photosynthesis
1 Selecting information...	from a line graph			dii				
	from a table					a, c		d
2 Presenting information...	as a line graph		f		b			c
	as a bar graph	e						
	as a pie chart						f	
3 Processing information...	as a ratio		d					
	as an average	c						
	as a percentage					d		e
	by general calculation [+ − × /]		e				c	
4 Planning & designing	Planning: aim, hypothesis, variables	a	c					
	Designing: apparatus, methods, replicates, controls	b	a, b	a, bi, di			a, di	b
5 Predicting & generalising	Predicting	g	h	bii				
	Generalising	d						
6 Concluding & explaining	Concluding	f		c, diii	ci			
	Explaining		g	bii	cii	b		a
7 Evaluating	Identifying source of error				ai		dii, e	
	Suggesting improvement				aii	e	b	
Total mark per question		**10**	**10**	**9**	**6**	**9**	**8**	**8**

1 Cell Biology

Question 1　Key Area 1.4 Measuring enzyme activity

Catalase is an enzyme found in the cells of living tissue where it speeds up the breakdown of hydrogen peroxide to form oxygen and water.

An experiment was set up to compare catalase activity in different living tissues.

Catalase was extracted from different types of living tissue as shown in the table below. The extracts were used to soak paper discs. The discs were dropped into **measuring cylinders** of hydrogen peroxide kept at 20 °C in a **water bath** and, as oxygen was released, the discs returned to the surface as shown in the diagram. The time taken for the disc to resurface was recorded for each tissue and the results shown in the table. The faster the disc returns to the surface, the faster the rate of the catalase reaction.

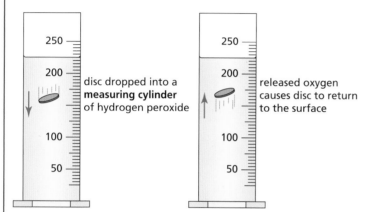

Type of living tissue		Time for disc to return to the surface (s)
Plant	Potato	64
	Carrot	44
	Apple	88
	Onion	40
Animal	Liver	4

a)　The experiments were carried out at 20 °C.
　　Identify **two** other variables which should be kept constant for each tissue tested to allow a valid comparison between the tissues. (2)

b)　Suggest a precaution which could be taken to increase the reliability of the results. (1)

c)　Calculate the average time for the discs to return to the surface for all the plant tissues. (1)

d)　Using the information given:
　　(i)　suggest a generalisation which could be made about the concentration of catalase present in plant compared with animal tissue (1)
　　(ii)　give a reason why this generalisation may not be reliable. (1)

e)　On a piece of graph paper, copy and complete the bar graph below by adding the vertical scale and the remaining bars to show the times taken for each disc to return to the surface. (2)

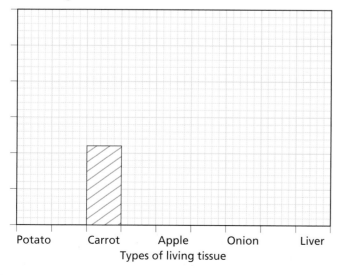

⇨
f) Using information given, write a conclusion appropriate to the aim of the experiment. (1)
g) Predict how the results for potato would differ if the experiment had been carried out at 30 °C. (1)

Question 2   Key Area 1.6 Using a respirometer

Six respirometers identical to the one shown in the diagram below were set up to show the effect of temperature on the rate of respiration in germinating pea seeds. Each respirometer was kept at a different temperature as shown in the table below.

The level of coloured liquid on the scale was brought to 0 at the start of each investigation and its movement up the scale was measured after 15 minutes. The results are shown in the table.

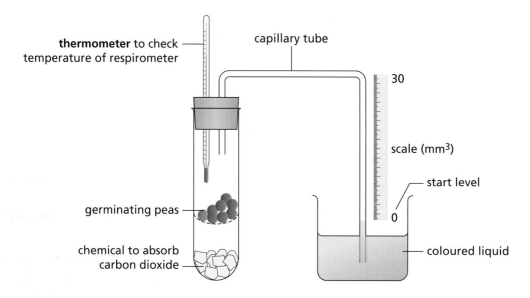

Respirometer	Temperature (°C)	Volume of oxygen consumed by germinating peas in 15 minutes (mm³)
1	5	2
2	10	8
3	15	16
4	20	22
5	30	28
6	50	4

a) Explain why the chemical to absorb carbon dioxide was included in each respirometer. (1)
b) Describe a suitable control for respirometer 2 which would ensure that the oxygen uptake by the peas was due to respiration. (1)
c) Give **two** variables which should be controlled to ensure that the results obtained were valid. (2)
d) Calculate the simplest whole number ratio of the oxygen consumed at 10 °C to that consumed at 30 °C. (1)
e) Calculate the oxygen uptake per hour by the peas in respirometer 3. (1)
f) On a piece of graph paper, copy and complete the grid below by adding a scale and label to the vertical axis and plotting a line graph of the results. (2)

⇨

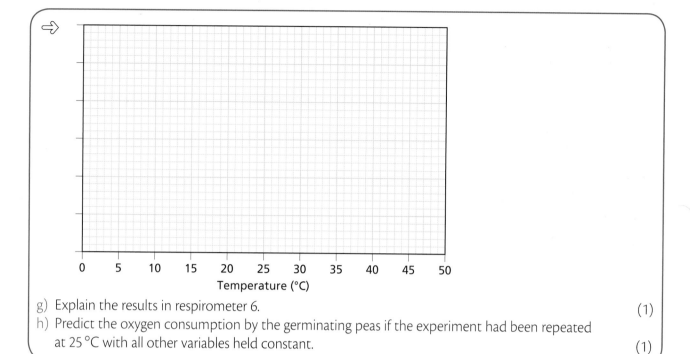

g) Explain the results in respirometer 6. (1)

h) Predict the oxygen consumption by the germinating peas if the experiment had been repeated at 25 °C with all other variables held constant. (1)

2 Multicellular Organisms

Question 3 Key Area 2.6 Using a potometer ?

The rate of transpiration in plants can be measured using a bubble potometer apparatus as shown in the diagram below.

As the leafy shoot transpires, water is drawn through the glass tube causing the air bubble trapped inside to move along the scale. The volume of water taken up by the shoot can be measured on the scale and the time taken can be measured on a **timer** or **stopwatch**. The potometer reservoir can be refilled as needed using water from a **syringe** or **dropping pipette**.

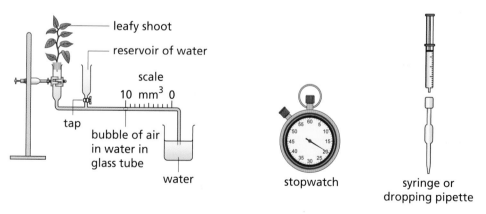

a) Using the information given, describe how the bubble of air can be returned to the zero mark on the scale following an experiment. (1)

b) The environmental factors listed below affect the rate of transpiration.

    wind speed   **air humidity**

(i) Choose **one** of these factors.

Describe one addition to the apparatus above which would allow an investigation into the effect of your chosen factor. (1)

(ii) 1 Predict how an **increase** in your chosen factor would affect the rate of transpiration in a leafy shoot. (1)

2 Explain why this prediction would happen. (1)

c) The diagram shows a weight potometer.

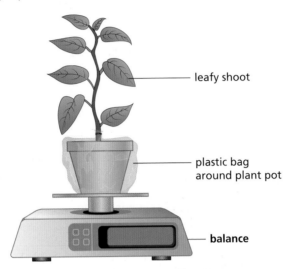

Explain how transpiration can be measured by this apparatus. (2)

d) Potometers were used to measure transpiration rates of leafy shoots of two different plant species L and M. Temperature was increased but all other conditions were kept constant. The results are shown on the graph below.

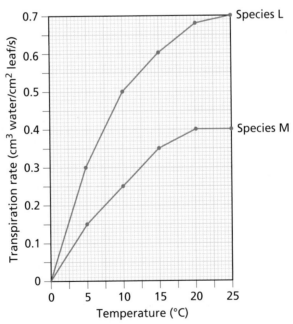

(i) Give **one** feature of the two leafy shoots which would have to be the same to allow valid comparison of their transpiration rates. (1)

(ii) Give the transpiration rate of **Species L** at 20 °C. (1)

(iii) Describe the evidence which supports the hypothesis that **Species M** is better adapted to life in hot, dry desert conditions than **Species L**. (1)

3 Life on Earth

Question 4   Key Area 3.2 Measuring an abiotic factor   ?

An investigation into differences in light intensity along a transect line extending from a grassland area into a woodland area was carried out.

A diagram of the transect line and the sampling stations used is shown below. At each sampling station, three readings were taken using a **light meter** held close to the ground and the results averaged. These are shown in the table below.

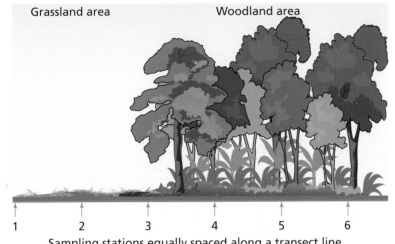

Sampling stations equally spaced along a transect line

Sampling station	Average light intensity (units)
1	50
2	30
3	10
4	5
5	2
6	2

a) (i) Describe **one** source of error in using a light meter which could lead to inaccurate readings.   (1)
   (ii) Describe how the results for average light intensity readings could be made more reliable.   (1)

b) On a piece of graph paper, copy and complete the line graph below to show how the average light intensity changes along the transect.   (2)

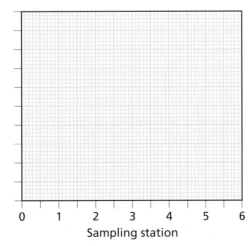

Sampling station

c) (i) Give **one** conclusion which can be drawn about the changes in light intensity along the transect.   (1)
   (ii) Explain how the changes in light intensity may affect the distribution of grass plants in the ecosystem.   (1)

Question 5   Key Area 3.2 Measuring the distribution of an organism

The table below shows the results of a study set up on a rocky shore to investigate the populations of seaweeds living there. A transect line was set out running from the high-tide mark at station 1 to the low-tide mark at station 10 as shown in the diagram below.

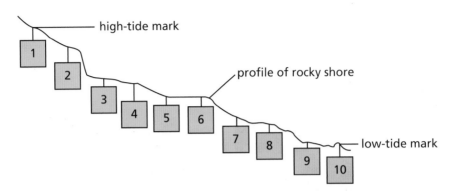

At each station, a quadrat was placed randomly and the percentage of the quadrat occupied by each of five seaweed species was recorded as shown in the table below.

| Station | Percentage cover by seaweed species | | | | |
	Kelp	Bladder wrack	Saw wrack	Channel wrack	Sea lettuce
1	0	0	0	0	0
2	0	0	0	20	0
3	0	0	30	50	0
4	0	0	40	20	0
5	0	0	90	10	0
6	0	30	60	10	0
7	0	100	0	0	0
8	0	90	0	0	10
9	80	5	0	0	5
10	100	0	0	0	0

Station 1 = high-tide mark; Station 10 = low-tide mark

a) Describe the distribution of saw wrack along the transect. (2)
b) Identify the seaweed species which spends the most time exposed to the atmosphere and justify your choice. (2)
c) Identify the station with the greatest seaweed biodiversity. (1)
d) Calculate the percentage difference in number of stations with channel wrack compared to the number with sea lettuce. (1)
e) Suggest an improvement to the sampling method at each station which could give more reliable estimates of the distribution of seaweed species along the transect. (1)

⇨

f) Using information from the table, copy and complete the pie chart on the right to show the percentages of the seaweed species present at station 6. (2)

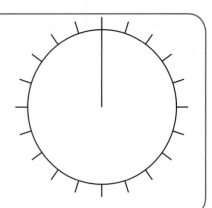

Question 6   Key Area 3.2 Using a transect line

An investigation was set up to measure the distribution of rye grass plants and woodlice in an area of vegetation. A transect line was set out as shown in the diagram below. At 10-metre intervals along the transect, a 0.25 m² quadrat was dropped randomly and the number of rye grass plants in each was recorded as shown in the table below. At each station, a pitfall trap was set overnight and the number of woodlice trapped in each by the next morning was counted.

Station	1	2	3	4	5	6	7	8
Number of rye grass plants present	3	2	4	2	1	2	0	0
Number of woodlice trapped	1	1	2	3	3	1	1	0

a) (i) Suggest why the investigators used a transect line as part of the sampling procedure instead of simply using eight randomly placed quadrats. (1)
   (ii) Give a reason why the quadrat was placed randomly at each station. (1)
   (iii) Explain why the pitfall traps were only left out for one night. (1)
b) It was suggested that the results of the investigation were not reliable.
   Describe **one** method of increasing the reliability of the sampling. (1)
c) Calculate the expected number of rye grass plants per square metre at station 1. (1)
d) Two abiotic factors were suggested which could be limiting the distribution of the rye grass.
   light intensity   **soil moisture content**
   Choose **one** abiotic factor from the list above.
   (i) Describe how you would make measurements of your chosen factor as part of this investigation. (2)
   (ii) Give **one** error which should be avoided when making measurements of this factor. (1)

Question 7   Key Area 3.3 Measuring the rate of photosynthesis

An experiment was set up to measure the effect of light intensity on the rate of photosynthesis in a water plant as shown in the diagram below.

The light intensity of the lamp was varied using a dimmer switch.

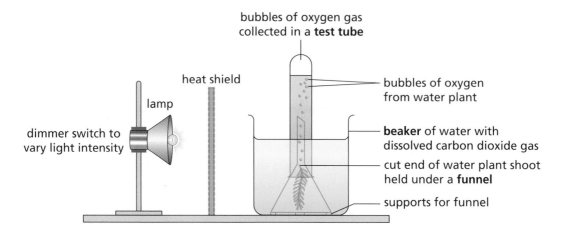

The rate of photosynthesis was measured by counting the number of bubbles of oxygen released per minute by a cut shoot of the water plant. The results are shown in the table below.

Light intensity (units)	Rate of photosynthesis (number of oxygen bubbles per minute)
2	4
4	20
6	40
8	90
10	90
12	90

a) Explain why measuring the rate of oxygen production by the water plant can be used as a measure of the rate of photosynthesis. (1)

b) (i) Explain why the heat shield is needed in the experiment. (1)

   (ii) Explain why the water in the beaker required dissolved carbon dioxide. (1)

c) On a piece of graph paper, copy and complete the grid on the next page by adding a scale and label to the horizontal axis and plotting a line graph of the results. (2)

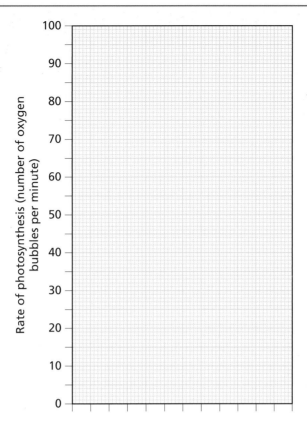

d) It was concluded that at 4 light intensity units, the rate of photosynthesis was being limited by the light intensity itself.

(i) Give evidence from the data which would support this conclusion. (1)

(ii) Suggest a factor which may be limiting the rate of photosynthesis at 10 units of light intensity. (1)

e) Calculate the percentage increase in rate of photosynthesis when the light intensity was increased from 6 to 8 units. (1)

4.3 Hints and tips

These are divided into the main skill areas but those directly related to **experimental skills** have been grouped together. Under **processing of information** we have shown how to tackle calculations involving percentage, ratio and average.

Tips on selecting information (from graphs)

- On graphs, the variable being investigated is on the *x*-axis and what is being measured is on the *y*-axis.
- Watch out for graphs with a double *y*-axis – these are tricky! The two *y*-axes often have different scales to increase the difficulty. You must take care to read the question and then the graph carefully to ensure that you are reading the correct *y*-axis.
- Work out the value of the smallest square on both scales before trying to read actual values from graphs.

- If you are asked to calculate an increase or decrease between points on a graph, you should use a ruler to help accuracy – draw pencil lines on the actual graph if this helps.
- When you are asked to describe a trend it is essential that you quote the values of the appropriate points and use the exact labels given on the axes in your answer. You must use the correct units in your description.

Tips on presenting information

- The most common question in this area requires students to present information that has been provided in a table as a graph – usually a line graph or sometimes a bar graph.
- Read the question to check whether a line graph or a bar graph is required – the question will usually tell you.
- Marks are given for providing scales, labelling the axes correctly and plotting the data points. Line graphs require points to be joined with straight lines using a ruler. Bar graphs need to have the bars drawn precisely using a ruler. The tops of the bars must be clearly seen.
- The graph labels should be identical to the table headings and units. Copy them exactly, leaving nothing out.
- You need to decide which variable is to be plotted on each axis. The data for the variable under investigation is placed in the left column of a data table and should be scaled on the x-axis. The right column in a data table provides the label and data for the y-axis. You will lose a mark if these are reversed.
- You must select suitable scales so that at least half of the graph grid provided is used, otherwise the mark will not be awarded. The value of the divisions on the scales you choose should allow you to plot all points accurately.
- Make sure that your scales extend beyond the highest data points.
- The scale must rise in regular steps. At National 5 level the examiners may test you on this by deliberately skipping one of these values that they have given you to plot in the table. Watch out for this deliberate change in the data, which is designed to check your care in plotting.
- Be careful to include one or both zeros at the origin if appropriate. It is acceptable for a scale to start with a value other than zero if this suits the data.
- Take great care to plot each point accurately using a cross or a dot and then connect them exactly using a ruler.
- Do not plot zero or connect the points back to the origin unless zero is actually included in the data table. If 0 is there you must plot it.
- When drawing a bar graph ensure that the bars are the same width and use a ruler, especially for the top of the bars. Remember to include a key if the data require it.
- If you make a mistake in a graph, another piece of graph paper is provided at the end of your exam paper.

Tips on processing information: tackling the common calculations

Percentages

Expressing a number as a %

The number required as a percentage is divided by the total and then multiplied by 100, as shown:

$$\frac{\text{number wanted as a \%}}{\text{total}} \times 100$$

Percentage change: increase or decrease

First, calculate the increase or decrease. Then, express this value as a percentage:

$$\frac{\text{change}}{\text{original starting value}} \times 100$$

Ratios

These questions usually require you to express the values given or being compared as a simple whole number ratio.

First you need to obtain the values for the ratio from the data provided in the table or graph. Take care that you present the ratio values in the order they are stated in the question. Then simplify them, first by dividing the larger number by the smaller one then dividing the smaller one by itself. However, if this does not give a whole number then you need to find another number that will divide into both of them. For example, 21:14 cannot be simplified by dividing 21 by 14 since this would not give a whole number. You must then look for another number to divide into both, in this case 7. This would simplify the ratio to 3:2, which cannot be simplified any further.

Averages

Add up the values provided and then divide the total by the number of values given.

Tips on experimental skills of planning and designing

Questions designed to test your skills in this area require you to discuss aspects such as **reliability**, **variables**, **validity**, **accuracy**, **controls**, measurements required, **sources of error** and suggested improvements. These questions often take the forms shown below.

- Give a reason for the experiment being repeated.
  - To improve the reliability of experiments and the results obtained, the experiment should be repeated.

- Give one precaution taken to ensure that the results would be valid.
  - To improve validity, only the one variable being investigated should be altered while the other variables should be kept constant.
- Describe a suitable control.
  - The control should be identical to the original experiment apart from the one factor being investigated.
- Explain why a control experiment was necessary.
  - A control experiment allows a comparison to be made and allows you to attribute any change or difference in the results to the factor or variable being altered.
- Why is it good experimental procedure to...?
  - If the effect of temperature on enzyme activity is being investigated, it is good practice to allow solutions of enzyme and substrate to reach the required temperature before mixing them to ensure that the reaction starts at the experimental temperature.
  - It is good experimental practice to use percentage change when you are comparing results because this allows a fair comparison to be made when the starting values in an investigation are different.

Precautions to minimise errors include washing apparatus such as beakers or syringes, or using different ones if the experiment involves different chemicals or concentrations. This prevents cross-contamination.

Questions regarding procedures that ask why the experiment was left for a certain time require you to state that this is to allow enough time for particular events to occur. These events could include the following:

- diffusion or absorption of substances into tissue
- growth taking place
- the effects of substances becoming visible
- a reaction occurring.

Experimental situations can often be modified to test different variables. If you are asked about this, think about how to alter different variables while keeping the original variable constant. For example, the *Cabomba* bubbler is often used to investigate variable light intensities. If light intensity was kept constant then the bubbler could be used to investigate temperature by exposing it to different temperatures while keeping the light intensity constant.

The context of scientific inquiry questions will usually be unfamiliar to you but the techniques and tips we have given should apply to most situations.

> **Hints & tips** ★
>
> *Examples of the variables that need to be controlled and kept constant to ensure results would be valid include temperature, pH, concentrations, mass, volume, length, number, surface area and type of tissue, depending on the actual experiment.*

> **Hints & tips** ★
>
> *If you are asked to describe a suitable control, make sure that you describe it in full.*

> **Hints & tips** ★
>
> *Watch out for the questions that refer to dry mass. Since the water content of tissues is variable and can change from day to day, dry mass is often used when comparing masses of tissues that are expected to change under experimental conditions.*

Answers to skills of scientific inquiry questions

4.1 An example: the *Cabomba* bubbler

Q1 To investigate the effects of changing light intensity on the rate of photosynthesis

Q2 Increasing light intensity will increase the rate of photosynthesis.

Q3 Produces bubbles of oxygen, which are a product of photosynthesis and can be easily seen and counted

Q4 Photosynthesis requires CO_2; the bicarbonate solution breaks down to give CO_2; in natural waters, respiration of living organisms would add CO_2 to the water.

Q5 It is a measure of oxygen produced; oxygen is a photosynthesis product.

Q6 Better to ensure that any trapped air is released first

Q7 Moving the lamp nearer to the plant increases the intensity of the light.

Q8 A stop clock or watch

Q9 Temperature of plant; concentration of CO_2 available; the size/species/type of plant

Q10 It increases the reliability of the results.

Q11 15 bubbles

Q12 50 cm

Q13 Axes with even scales and labels with units; accurate plotting and connecting with straight lines

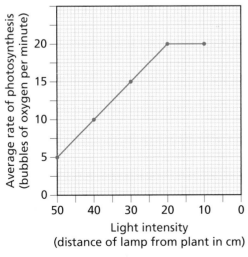

Q14 4:1

Q15 14 bubbles per minute

Q16 300%

Q17 At 5 cm, no increase in the rate of photosynthesis compared with 10 cm; at 60 cm, lower than that at 50 cm

Q18 Expect increased rate of photosynthesis/ number of bubbles of oxygen produced

Q19 You would expect an increase at higher intensities because glucose is another product of photosynthesis.

Q20 As light intensity increases, the rate of photosynthesis increases up to 20 cm, after which increases in light intensity cause no further increase in rate of photosynthesis.

Q21 Light intensity is a limiting factor, so it should limit the rate at low intensities; at high intensities other factors could start to limit photosynthesis.

Q22 The bubbles might not be pure oxygen.

Q23 It does matter and the bubbles are probably not the same size. The gas could be collected in an inverted measuring cylinder to measure the volume of oxygen produced.

Q24 By ensuring that distance is measured accurately to the glass of the bulb **OR** using a light meter. Use a dimmer control on the light.

Q25 Other light does matter. The experiment could be carried out in a darkened room.

Q26 Temperature and CO_2 concentration

Q27 Probably not reliable enough – more sets of results/data needed.

Q28 Using a dimmer switch and light meter to control and measure light intensity; using a plate glass heat shield to cut down heating effects of lamp; collecting oxygen in a measuring cylinder; repeating the experiment at each light intensity and averaging results; monitoring temperature of water using a thermometer

Q29 Temperature and CO_2 concentration

Q30 By keeping light intensity constant, it would be possible to alter either temperature or CO_2 concentration. ⇨

4.2 Skill by skill questions

Q1 Measuring enzyme activity

a) volume/concentration of hydrogen peroxide

OR size, diameter/thickness of disc

OR mass/volume of extract

[any 2, 1 mark each]

b) repeat the investigation/use more discs of each tissue

c) 48 s

d) (i) catalase is more concentrated in animal tissues

(ii) not enough different animal tissues tested

e) evenly stepped scale and label on vertical axis = 1

accurate bars with straight line tops = 1

f) different tissues have different catalase levels/activities

g) would return to surface in less than 64 s

Q2 Using a respirometer

a) so that the movement of the coloured liquid would represent the oxygen consumed by the peas only

b) an identical respirometer but containing an equal mass of dead peas

c) equal mass of chemical to absorb carbon dioxide

OR equal mass of germinating peas

OR equal volume of air in the respirometer

[any 2, 1 mark each]

d) 2:7

e) 64 mm³

f) evenly stepped scale and label on vertical axis = 1

accurate points joined by straight lines = 1

g) rate of respiration drops because enzymes which control respiration become denatured

h) 25 mm³ in 15 minutes

Q3 Using a potometer

a) open the tap on the reservoir and close again when bubble reaches 0

b) (i) wind speed: add a fan/hairdrier on a cold setting to create moving air

OR air humidity: use a plastic bag to cover the shoot and raise humidity

(ii) 1 wind speed increase would increase rate of transpiration

OR air humidity increase would decrease rate of transpiration

2 wind speed increase would increase the rate of evaporation of water

OR air humidity increase would decrease the rate of evaporation of water

c) weigh the plant on the balance = 1

allow to transpire for a set period then re-weigh (to measure the water loss in the period) = 1

d) (i) their surface area

(ii) 0.68 cm³ water per cm² leaf per s

(iii) loses less water than Species L as the temperature increases

Q4 Measuring an abiotic factor

a) (i) shading to light sensor by investigator

(ii) take more readings at each station

b) evenly stepped scale and label on vertical axis = 1

accurate points joined by straight lines = 1

c) (i) as the transect moves towards and into the wood, the light intensity decreases

(ii) lower light intensity at ground level reduces photosynthesis in grasses and so reduces their abundance

Q5 Measuring the distribution of an organism

a) saw wrack is absent until station 3 and then increases until station 5 = 1

then decreases at station 6 and then is absent through to station 10 = 1

b) channel wrack = 1

which is found highest up the shore (and so spends least time covered by the tide) = 1

c) station 6

d) channel wrack has 30% more stations than sea lettuce
OR sea lettuce has 30% fewer stations than channel wrack

e) use more quadrats at each station and average the results

f) correct sectors = 1
correct labels = 1

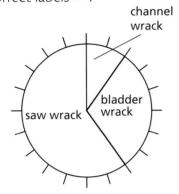

Q6 Using a transect line

a) (i) because the communities in the habitat are variable/are stratified/change along a line **OR** transect makes the sampling more representative

(ii) to eliminate bias from the sample **OR** to prevent unrepresentative sampling

(iii) if traps are left too long, predators may eat animals in the sample

b) use more/bigger quadrats at each station

c) 12 plants per square metre

d) (i) light intensity: use a light meter/photometer = 1
at each station = 1
OR soil moisture content: use a moisture meter (and probe) = 1
at each station = 1

(ii) light intensity meter: avoid shading the light sensor when taking readings
OR soil moisture meter: wipe/clean probe between readings

Q7 Measuring the rate of photosynthesis

a) because oxygen is a product of photosynthesis

b) (i) to prevent heat from the lamp affecting the rate of photosynthesis so ensuring that only one variable is altered

(ii) carbon dioxide needed because it is a raw material for photosynthesis

c) evenly stepped scale and label on horizontal axis = 1
accurate points joined by straight lines = 1

d) (i) when the intensity was raised to 6 units, the rate of photosynthesis increased

(ii) temperature **OR** carbon dioxide concentration

e) 125 % increase

Key words

Accuracy – measure of the degree of precision of a measurement

Balance – instrument for measuring the weight or mass of materials

Beaker – container to hold and pour liquids for use in experiments

Control – technique which can help to ensure that experimental results are caused by the factor under investigation

Dropping pipette – instrument for delivering small measured volumes of liquid

Funnel – device for channelling liquids or gas bubbles in an experiment

Measuring cylinder – container to hold and measure volumes of liquid

Meter – instrument for measuring abiotic factors such as light or moisture

Reliability – measure of confidence in the results of an experiment

Source of error – a factor or feature of experimental procedure which can lead to mistakes

Stopwatch – instrument for measuring time in an experiment

Syringe – instrument for delivering small measured volumes of liquid

Test tube – glass container for holding liquid mixtures or trapping gases in experiments

Thermometer – instrument for measuring temperature in an experimental situation

Validity – measure of the fairness of an experiment

Variable – factor in an experiment which is changeable or can change

Water bath – trough of thermostatically heated water to provide steady controlled temperatures

Your Assignment

The assignment is based on an investigative research task that you have carried out in class time. You choose the topic to be studied and then investigate and research it. The assignment must be an individually produced piece of work, although experimental work and fieldwork can be done in small groups and data shared.

The assignment assesses the application of skills of scientific inquiry and some related biology knowledge and understanding that you develop as you work through the course.

You will have to write up the work in the form of a report under controlled assessment conditions. During your write-up you will have access to notes made during the research stage but **not** a draft copy of any part of your final assignment report.

The report will be marked out of 20 marks with 17 of the marks being for scientific inquiry skills and 3 marks for the demonstration and application of knowledge of background biology.

The report is marked by the SQA and contributes 20% to the overall grade for your course.

The assignment is supervised by your teachers who should supply you with the *Instructions for Candidates* document published by SQA. Your assignment has **two** main stages as shown in the table below. The stages can be split into tasks as shown.

Stage		Task		Recommended timing	Supervision level
1	Research	A	Selecting a topic; devising an aim; planning experiments or fieldwork; summarising the background biology	6 hours and 30 minutes	Some supervision and control
		B	Carrying out experimental work or fieldwork; collecting numerical results; drawing conclusions; evaluating your work		
		C	Selecting information from existing sources to compare to experimental or fieldwork results		
2	Report	D	Writing a report	1 hour and 30 minutes	High degree of supervision and control

Outline of the stages in the assignment

1 Research stage

Task A

Selecting a topic and a title

Although you can choose a topic yourself, your teacher will probably support you with some ideas to choose from. Make sure you choose something relevant that you are interested in and that you understand. It might be a good idea to think about using one of the experimental techniques which are needed for the course anyway. The table below reminds you about the techniques and, in each case, some possible assignment topics they relate to. **You should note that these are topics, not assignment aims**. Your title should be informative and might indicate the topic area but your aim will need to be more specific.

Technique	Page	Assignment topic areas
Measuring enzyme activity		Pectinase and apple juice production Biological washing powders
Using a respirometer		Comparing respiration rates Industrial uses of fermenters
Using a potometer		Adaptations of desert plants Effects of climate change on the water cycle
Measuring abiotic factors		Effects of climate change on species Effects of fertilisers on plant growth Fertilisers and pollution Factors affecting seed germination
Measuring the distribution of a species		Plant distribution and light Zonation in seaweeds Plant succession on sand dunes Plant succession around freshwater lochs
Using a transect line		
Measuring the rate of photosynthesis		Factors affecting crop production Limiting factors in photosynthesis

There are many other possible experimental techniques you could use leading to many other assignment topics.

Devising an aim

Ensure that you have a clear and specific aim for the assignment. For example, an acceptable aim might be 'to investigate the effect of temperature on enzyme activity'. You will need to state your aim clearly in your report.

Summarising relevant background biology

The background biology which you give must be relevant to your aim. To score full marks, you must give at least three relevant points. Ideally, you should keep your underlying biology material in a separate section of your report. It is not necessary to give references for this section.

Planning your investigation

Think about what you will do. What apparatus will you need? What steps will you take? What measurements will you make? How will you collect and record your results data? You need to think about controls and reliability. You will have to summarise your methods in your report.

Task B

Carrying out experimental work or fieldwork

You must carry out experimental work or fieldwork safely to generate data relevant to your aim. This will involve carrying out your planned experiment under the supervision of your teacher. You are able to work actively with a partner or small group and share results data.

Collecting numerical data

You must record your experimental data in numerical form. It is sensible to design and draw some table grids for your results before you start carrying out your work. You must have your own copy of the raw results data. During your report write-up, you must process your data by, for example, calculating mean values and presenting the data as a line graph or bar graph as appropriate to the data itself.

Drawing a conclusion

You must draw a conclusion. This has to be relevant to your aim and must be supported by the data and other information you have collected. If the pre-existing source of data does not support your own data, you must state this in your report and say why you think this is. Note that the mark for concluding cannot be awarded if you have not stated an aim in the first place.

Evaluating

In your report, you must evaluate any experimental work or fieldwork you carried out. The evaluation can focus on reliability, validity and accuracy of measurements. You should suggest an improvement which could minimise the effect of any error.

Task C

Selecting information from existing sources

You must research your aim and select information from books, journals or the internet. Your teacher will probably supply you with a number of sources from which you can choose.

Comparing your own data with an existing source

The information you find has to be compared with the data you generated during your experimental work or fieldwork. If a comparison is not possible, you must say so in your report and give an explanation of why not.

You must give a reference to the source of your pre-existing data in your report as shown in the table below.

Source	Reference
Website	Full URL for the page or pages i.e. the URL 'www.sqa.org.uk' is not acceptable, but the following is an acceptable reference: http://www.sqa.org.uk/files_ccc/BiologyN5CAT.pdf
Journal	Title, author, journal title, volume and page number
Book	Title, author, page number and either edition or ISBN

2 Report stage

Task D

Writing up your assignment report

You will have to write up your report under controlled conditions, with access to notes only. The maximum time allowed for this stage is 1 hour and 30 minutes. Your report should have the sections or features shown in the table below. The table also shows the allocation of marks.

You must not have a draft report or a draft of any section to work from. It is expected that you will write your report with the assistance only of your notes.

Section of assignment report (mark allocation)		Notes	Marks
1	Aim (1)	Clearly describes the purpose of the investigation	1
2	Underlying biology (3)	With descriptions and explanations at N5 level	3
3	Data collection and handling (6)	Summary of the method used to collect experimental or fieldwork data	1
		Raw data from experiments or fieldwork included	1
		Data presented as a table	1
		Data processed e.g. as mean values	1
		Relevant data from pre-existing sources presented	1
		Reference to pre-existing sources given	1
4	Graphical presentation (4)	Appropriate format: either line graph or bar graph	1
		Suitable scales	1
		Scales given labels and units	1
		Data plotted accurately using a ruler	1
5	Analysis (1)	A valid comparison of experimental or fieldwork data with the selected pre-existing data	1
6	Conclusion (1)	A valid conclusion which is related to the aim and supported by evidence	1
7	Evaluation (2)	Identification of a source of error which focuses on reliability, accuracy or precision	1
		A suggested improvement to minimise the effect of the identified source of error	1
8	Structure (2)	An informative title	1
		A clear and concise report which flows naturally	1
Total marks			**20**

General exam revision: 20 top tips

These are very general tips and would apply to all your exams.

1 Start revising in good time.

Don't leave it until the last minute – this will make you panic and it will be impossible to learn. Make a revision timetable that counts down the weeks to go.

2 Work to a study plan.

Set up sessions of work spread through the weeks ahead. Make sure each session has a focus and a clear purpose. What will you study, when and why?

3 Make sure you know exactly when your exams are.

Get your exam dates from the SQA website and use the timetable builder tool to make up your own exam timetable. You will also get a personalised timetable from your school but this might not be until close to the exam period.

4 Make sure that you know the topics that make up each course.

Studying is easier if material is in chunks – why not use the SQA Key Areas? Ask your teacher for help on this if you are not sure.

5 Break the chunks up into even smaller bits.

The small chunks should be easier to cope with. Remember that they fit together to make larger ideas. Even the process of chunking down will help!

6 Ask yourself these key questions for each course.

Are all topics compulsory or are there choices? Which topics seem to come up time and time again? Which topics are your strongest and which are your weakest?

7 Make sure you know what to expect in the exam.

How is the paper structured? How much time is there for each question? What types of question are involved – multiple choice, structured response, extended response?

8 There is no substitute for past papers – they are simply essential!

The specimen and past papers for all courses are on the SQA website – look for the past paper finder and download as PDF files. There are answers and mark schemes too.

9 Use study methods that work well for you.

People study and learn in different ways. Reading and looking at diagrams suits some people. Others prefer to listen and hear material – what about reading out loud or getting a friend or family member to do this for you? You could also record and play back material.

10 Make sure the information reaches your long-term memory.

There are only three ways of doing this:
- practice – e.g. rehearsal, repeating
- organisation – e.g. making drawings, lists, diagrams, tables, memory aids
- elaborating – e.g. winding the material into a story or an imagined journey

11 Learn actively.

Most people prefer to learn actively – for example, making notes, highlighting, redrawing and redrafting, making up memory aids, writing past paper answers.

12 Be an expert.

Be sure to have a few areas in which you feel you are an expert. This often works because at least some of them will come up, which can boost confidence.

13 Try some visual methods.

Use symbols, diagrams, charts, flashcards, post-it notes etc. The brain takes in chunked images more easily than loads of text.

14 Remember – practice makes perfect.

Work on difficult areas again and again. Look and read – then test yourself. You cannot do this too much.

15 Try past papers against the clock.

Practise writing answers in a set time. As a rough guide, you should be able to score a mark per minute.

16 Collaborate with friends.

Test each other and talk about the material – this can really help. Two brains are better than one! It is amazing how talking about a problem can help you solve it.

17 Know your weaknesses.

Ask your teacher for help to identify what you don't know. If you are having trouble, it is probably with a difficult topic so your teacher will already be aware of this – most students will find it tough.

18 Have your materials organised and ready.

Know what is needed for each exam. Do you need a calculator or a ruler? Should you have pencils as well as pens? Will you need water or paper tissues?

19 Make full use of school resources.

Are there study classes available? Is the library open? When is the best time to ask for extra help? Can you borrow textbooks, study guides, past papers etc.? Is school open for Easter revision?

20 Keep fit and healthy!

Mix study with relaxation, drink plenty of water, eat sensibly, and get fresh air and exercise – all these things will help more than you could imagine. If you are tired, sluggish or dehydrated, it is difficult to see how concentration is even possible.

National 5 Biology: 20 top exam tips

These tips apply specifically to National 5 Biology. Remember that your assignment is worth 20% of your grade; the other 80% comes from the 100 marks in the examination.

Section 1: Multiple choice (25 marks)

1 Do not spend more than *40 minutes* on this section.
2 Answer on the grid provided.
3 *Do not leave blanks*. Complete the grid for each question as you work through.
4 Try to answer each question in your head *without* looking at the options. If your answer is there, you are home and dry!
5 If you are not certain, choose the answer that seemed most attractive on *first* reading the options. If you are forced to guess, try to eliminate some options before making your guess. If you can eliminate three, you are left with the correct answer even if you do not recognise it!

Section 2: Structured and extended response (75 marks)

1 Spend about *110 minutes* on this section.
2 Answer on the question paper. Try to write neatly and keep your answers on the support lines if possible; these are designed to take the full answer.
3 Another clue to answer length is the mark allocation. Most questions are restricted to 1 mark and the answer can be quite short; if there are 2–5 marks available, your answer will need to have the detail to reflect this.
4 The questions are usually laid out in course sequence but remember that some questions are *designed* to cover more than one Key Area.
5 Grade C (less demanding) questions usually start with 'State', 'Give' or 'Name'.
6 Grade A (more demanding) questions begin with 'Explain' and 'Describe' and are likely to have more than one part to the full answer.
7 Abbreviations like DNA and ATP are fine.
8 Don't worry that some questions are in unfamiliar contexts. This is deliberate. Just keep calm and read the questions carefully.
9 If a question contains a choice, be sure to spend enough time making the right choice.

10 Remember to *use values from the graph* if you are asked to do so.

11 Draw graphs using a ruler and use the data table headings for the axes labels.

12 Look out for graphs with two *y*-axes. These need extra concentration as they can easily lead to mistakes.

13 Answers to calculations will not usually have more than two decimal places.

14 If there is a space for calculation given, it is very likely that you will need to use it.

15 Do not leave blanks. Have a go, using the language in the question if you can.

Glossary

The terms included here appear in the SQA Assessment Specification for National 5 Biology. They are defined here in the context of National 5. The Key Area in which a term first appears is given in brackets after each term. If the term first appears in the Skills of Scientific Inquiry section, the reference is given in brackets as SSI.

Where a term has an unusual singular or plural, this is given in brackets with the definition.

You could make flashcards with the term on one side and the meaning on the other – a great resource for revision!

A, G, T, C (1.3)   letters that represent the names of bases of DNA and mRNA (adenine, guanine, thymine, cytosine)

Abiotic factor (3.2)   a physical factor, such as temperature or light intensity, that affects ecosystems

Absorption (2.7)   the taking up of molecules by cells

Accuracy (SSI)   measure of the degree of precision of a measurement

Active site (1.4)   position on the surface of an enzyme molecule to which specific substrate molecules can bind

Active transport (1.2)   transport of molecules against their concentration gradient

Adaptation (3.6)   feature of an organism that helps it to survive

Aerobic (1.6)   in the presence of oxygen

Aerobic respiration (1.1)   release of energy from food by a cell using oxygen

Algal bloom (3.5)   rapid increase of algal numbers in water

Allele (2.4)   a form of a gene

Alveoli (2.7)   tiny sacs in lungs that form the gas exchange surface (*sing.* alveolus)

Amino acid (1.3)   a building block of a protein molecule

Anther (2.3)   organ within a flower that produces pollen grains

Antibody (1.4)   protein that is involved in defence in animals

Aorta (2.6)   main artery that carries oxygenated blood away from the heart in mammals

Artery (2.6)   general name for a blood vessel that carries blood away from the heart

ATP (1.6)   adenosine triphosphate; a substance that transfers chemical energy in cells

Atria (2.6)   upper chambers of the heart, which receive blood from veins (*sing.* atrium)

Bacterial cell (1.1)   the tiny individual cell of a bacterium

Balance (SSI)   instrument for measuring the weight or mass of materials

Bases (1.3)   form the genetic code of DNA and mRNA

Beaker (SSI)   container to hold and pour liquids for use in experiments

Biconcave (2.6)   shape of a red blood cell describing dimples which increase the surface area of the cell

Bioaccumulation (3.5)   the build-up of toxic substances in living organisms

Biodiversity (3.1)   refers to the number and abundance of species

Biological control (3.5)   natural control of pests using natural predators, parasites etc.

Biotic factor (3.2)   factor related to the biological aspects of an ecosystem such as predation or competition

Bond (1.3)   chemical link between atoms in a molecule

Brain (2.2a)   organ of the central nervous system of mammals where vital functions are coordinated

Capillaries (2.6)   tiny blood vessels with walls one cell thick where exchange of materials occurs

Capillary network (2.7)   branching system of tiny blood vessels providing a large surface area for absorption

Carbohydrate (3.3)   a substance such as sugar, starch, cellulose or glycogen, containing the elements carbon, hydrogen and oxygen

Carbon fixation stage (3.3)   second stage in photosynthesis in which ATP, hydrogen and carbon dioxide are involved in the production of sugar

Carnivore (3.1)   consumer which eats other animals only

Catalyst (1.4)   a substance that speeds up a chemical reaction by reducing the energy required to start it

Cell (1.1)   a basic unit of life

Cell membrane (1.1)   selectively permeable layer consisting of phospholipid and protein, enclosing the cell cytoplasm and controlling the entry and exit of materials

Cell wall (1.1)   supports cells and prevents them from bursting; plant, fungal and bacterial cell walls have different structures and chemical compositions

Cellulose (1.1)   structural carbohydrate of which plant cell walls are composed

Central nervous system (CNS) (2.2a)   part of the nervous system made up of the brain and spinal cord

Cerebellum (2.2a)   part of the brain that controls balance and coordination of movement

Cerebrum (2.2a)   large folded part of the brain that controls conscious thoughts, reasoning, memory and emotion

Chlorophyll (3.3)   green pigment in chloroplasts that absorbs light energy for the process of photosynthesis

Chloroplast (1.1)   organelle containing chlorophyll; the site of photosynthesis

Chromatid (2.1)   replicated copy of a chromosome visible during cell division

Chromosome (2.1) structure containing hereditary material; composed of DNA that codes for all the characteristics of an organism

Chromosome complement (2.1) the characteristic number of chromosomes in a typical cell of a species

Community (3.1) all the organisms living in a habitat

Companion cell (2.5) living phloem cell with a nucleus found associated with sieve tube cells

Competition (3.1) interaction between organisms seeking the same limited resources

Complementary (1.3) fitting together as applied to DNA base pairing

Concentration gradient (1.2) difference in concentration between two solutions, cells or solutions and cells

Consumer (3.1) animal which obtains energy by eating other organisms

Continuous variation (2.4) variation which shows a range of values between extremes

Control (SSI) technique which can help to ensure that experimental results are caused by the factor under investigation

Coronary arteries (2.6) blood vessels that carry oxygenated blood to the heart tissues

Cytoplasm (1.1) jelly-like liquid which contains cell organelles and is the site of many chemical reactions

Decomposers (3.4) organisms such as bacteria and fungi responsible for the breakdown of dead organic material

Degradation (1.4) enzyme-controlled reaction in which a complex substrate is broken down into simple products

Denaturation (1.4) change in the shape of molecules of protein such as enzymes, resulting in them becoming non-functional

Diffusion (1.2) passive movement of molecules from an area of high concentration to an area of lower concentration

Digestion (2.7) breakdown of large, insoluble food molecules into smaller, soluble ones

Diploid (2.1) describes a cell containing two sets of chromosomes

Discrete (2.4) describes variation that is clear-cut and can be categorised into distinct groups

DNA (1.1) deoxyribonucleic acid; substance in chromosomes that carries the genetic code of an organism

Dominant (2.4) form of a gene that is expressed in the phenotype, whether homozygous or heterozygous

Double-stranded helix (1.3) describes the spiral ladder shape of DNA molecules

Dropping pipette (SSI) instrument for delivering small measured volumes of liquid

Ecosystem (3.1) natural biological unit composed of habitats, populations and communities

Effectors (2.2a) muscles or glands which make the response in a reflex arc

Egg cell (2.3)   female gamete produced by ovaries in animals

Electrical impulses (2.2a)   method of carrying messages along neurons

Embryo (2.3)   structure produced when a zygote divides repeatedly

Endocrine gland (2.2b)   gland that produces and releases a hormone directly into the blood

Enzyme (1.4)   protein produced by living cells that acts as a biological catalyst

Equator (2.1)   middle position of a spindle where chromosomes are attached during mitosis

Ethanol (1.6)   alcohol produced as a result of fermentation of sugars by yeast

F_1 **(2.4)**   first generation of a monohybrid cross

F_2 **(2.4)**   second generation of a monohybrid cross

Family tree (2.4)   diagram that shows the inheritance of a genetic condition in a family

Fatty acids (2.7)   products of fat digestion absorbed into lacteals

Fermentation (1.6)   a type of respiration without oxygen

Fertilisation (2.3)   the fusion of the nuclei of haploid gametes

Fertiliser (3.5)   chemical added to the soil to improve plant growth or crop yield

Food chain (3.4)   simple set of feeding links showing how energy passes from a producer to consumers

Fungal cell (1.1)   individual cell of a fungus

Funnel (SSI)   device for channelling liquids or gas bubbles in an experiment

Gamete (2.3)   sex cell containing the haploid chromosome complement

Gene (1.3)   small section of DNA that codes for the production of a specific protein

Genetic code (1.3)   code formed by the sequence of the bases in DNA that determines an organism's characteristics

Genetic engineering (1.5)   the artificial transfer of genetic information from one donor cell or organism to another

Genetically modified (GM) (1.5)   term used to describe a cell or organism that has had its genetic information altered, usually by adding a gene from another organism

Genotype (2.4)   the alleles that an organism has for a particular characteristic, usually written as symbols

Gland (2.2a)   group of cells which can synthesise and release a substance such as a hormone

Glucagon (2.2b)   hormone produced by the pancreas, responsible for triggering the conversion of glycogen into glucose in the liver

Glucose (1.6)   simple sugar used as a respiratory substrate for the production of ATP

Glycerol (2.7)   product of fat digestion absorbed into lacteals

Glycogen (2.2b)   animal storage carbohydrate located in the liver and muscle tissues

Grazing (3.2)   method of feeding by which a herbivore eats grasses and herbs

Guard cells (2.5)   found on either side of a stoma; they control gas exchange in leaves by controlling opening and closing of the stoma

Habitat (3.1)   the place where an organism lives

Haemoglobin (2.6)   pigment in red blood cells that transports oxygen as oxyhaemoglobin

Haploid (2.3)   describes a cell, usually a gamete, containing one set of chromosomes

Heart (2.6)   muscular organ that pumps blood around the body

Heart valves (2.6)   structures in the heart which prevent backflow of blood

Herbivore (3.1)   consumer which eats plants only

Heterozygous (2.4)   describes a genotype in which the two alleles for the characteristic are different

Homozygous (2.4)   describes a genotype in which the two alleles for the characteristic are the same

Hormone (1.4)   protein released by an endocrine gland into the blood to act as a chemical messenger

Host bacterial cell (1.5)   a bacterial cell that receives genetic material from a source chromosome

Immune system (2.6)   organs and tissues which resist infections

Indicator species (3.2)   organisms that by their presence, abundance or absence give information such as level of pollution in the environment

Insulin (2.2b)   hormone produced by the pancreas that triggers the conversion of glucose into glycogen in the liver

Inter neuron (2.2a)   nerve cell that transmits electrical impulses from sensory neurons to motor neurons within the CNS

Interspecific competition (3.1)   competition between organisms of two different species for a common resource

Intraspecific competition (3.1)   competition between organisms within the same species

Iodine solution (3.3)   brown-coloured solution that turns blue-black with starch

Isolation barrier (3.6)   barrier that prevents interbreeding and gene exchange between sub-populations of a species

Lactate (1.6)   substance produced during fermentation in animals and responsible for muscle fatigue

Lacteal (2.7)   fine tube in a villus providing a large surface area for the absorption of fatty acids and glycerol

Leaf epidermis (2.5)   skin-like tissue forming the upper and lower covering of leaves

Light reactions (3.3)   first stage in photosynthesis producing hydrogen and ATP required in the carbon fixation stage

Lignin (2.5)   carbohydrate material lining the xylem vessels and providing strength and support

Limiting factor (3.3)   a variable that, when in short supply, can limit the rate of a chemical reaction such as photosynthesis

Liver (2.2b)   large organ, beside the stomach, with many important functions including a role in blood glucose control

Lungs (2.7)   organs responsible for gas exchange

Lymphocytes (2.6)   white blood cells, some of which can produce antibodies to destroy specific pathogens

Measuring cylinder (SSI)   container to hold and measure volumes of liquid

Medulla (2.2a)   part of the brain controlling breathing, heart rate and peristalsis

Messenger RNA (mRNA) (1.3)   substance that carries a complementary copy of the genetic code from DNA to the ribosomes

Meter (SSI)   instrument for measuring abiotic factors such as light or moisture

Microscope (1.1)   an instrument for magnifying objects such as cells for study

Minerals (2.5)   nutrient ions essential for healthy growth

Mitochondrion (1.1)   organelle that is the site of aerobic respiration and ATP production in cells (*pl.* mitochondria)

Mitosis (2.1)   division of the nucleus of a cell that leads to the production of two genetically identical diploid daughter nuclei

Modified plasmid (1.5)   bacterial plasmid into which a required gene for a desirable characteristic has been inserted

Monohybrid cross (2.4)   cross set up to study the inheritance of contrasting characteristics caused by alleles of the same gene

Motile (2.3)   able to move under its own power

Motor neuron (2.2a)   nerve cell that carries electrical impulses from the CNS to effectors such as muscles or glands

Multicellular (1.1)   having many cells

Mutation (3.6)   a random and spontaneous change in the structure of a gene, chromosome or number of chromosomes; only source of new alleles

Natural selection (3.6)   process which ensures the best adapted individuals survive

Nerves (2.2a)   bundles of neurons that connect receptors to the CNS and the CNS to the effectors

Neuron (2.2a)   nerve cell that is specialised to transmit electrical impulses

Niche (3.1)   the role an organism has within its community in an ecosystem

Nitrates (3.5)   nutrients absorbed from soil by plants to produce amino acids

Nucleus (1.1)   organelle which controls cell activities and contains the genetic information of the organism (*pl.* nuclei)

Omnivore (3.1)   consumer which eats both plants and animals

Optimum (1.4)   conditions such as temperature and pH at which an enzyme is most active

Organ (2.1)   a group of different tissues that work together to carry out a particular function, e.g. heart and lungs

Organelle (1.1)   membrane-bound compartment with a specific function in animal, plant and fungal cells

Osmosis (1.2)   movement of water molecules from an area of high to an area of lower water concentration through a selectively permeable membrane

Ovaries (2.3)   female sex organs (*sing.* ovary)

Ovule (2.3)   structure containing a female gamete, produced by ovaries in plants

Oxyhaemoglobin (2.6)   substance produced when oxygen combines with haemoglobin

Paired statement key (3.2)   used to identify and name an unknown living organism

Palisade mesophyll (2.5)   main photosynthetic tissue in leaf containing many chloroplasts

Pancreas (2.2b)   organ responsible for the production of digestive enzymes and the hormones insulin and glucagon

Parental generation (P) (2.4)   parents of a monohybrid cross

Passive transport (1.2)   movement of molecules down a concentration gradient without the need for additional energy, e.g. diffusion and osmosis

Pathogen (2.6)   a disease-causing organism such as certain bacteria, viruses and fungi

Pesticides (3.5)   general name for chemicals used to kill organisms that damage or feed on crop plants

Petri dish (1.1)   shallow transparent container used for growing bacteria and carrying out other experiments

Phagocyte (2.6)   white blood cell which can engulf and digest pathogens during phagocytosis

Phenotype (2.4)   the visible characteristics of an organism that occur as a result of its genes

Phloem (2.5)   tissue in plants that transports sugar

Phospholipid (1.2)   component of cell membranes

Photosynthesis (1.1)   process carried out by green plants to make their own food using light energy

Pitfall trap (3.2)   sampling technique used to trap animals living on the soil surface or in leaf litter

Plasma (2.6)   liquid part of blood in which cells are suspended

Plasmid (1.1)   circular genetic material present in bacterial cells and used in genetic engineering or modification

Plasmolysed (1.2)   description of a plant cell in which the vacuole has shrunk and the membrane has pulled away from the wall due to water loss

Poles (2.1)   opposite ends of a spindle to which chromatids migrate during mitosis

Pollen grain (2.3)   structure produced in the anthers of a flower that contains the male gamete

Polygenic (2.4)   inheritance determined by the interaction of several genes acting together

Population (3.1)   the number of organisms of the same species living in a defined area

Potometer (2.5)   apparatus used to measure transpiration rate in leafy plants

Predation (3.1)   obtaining food by hunting and killing prey organisms

Predator (3.1)   animal which obtains food by hunting and killing prey organisms

Prey (3.1)   animal which is hunted and eaten by a predator

Producer (3.1)   green plant which makes food by photosynthesis

Product (1.4)   substance made by an enzyme-catalysed reaction

Protein (1.1)   substance composed of chains of amino acids and containing the elements carbon, hydrogen, oxygen and nitrogen

Pulmonary artery (2.6)   artery carrying deoxygenated blood from the heart to the lungs

Pulmonary vein (2.6)   vein carrying oxygenated blood to the heart from the lungs

Punnett square diagram (2.4)   a grid diagram which shows the predicted genotypes and proportions of offspring from a cross

Pyramid of energy (3.4) diagram that shows the relative quantities of energy at each level in a food chain

Pyramid of numbers (3.4) diagram that shows the relative numbers of organisms at each level in a food chain

Pyruvate (1.6) substance produced by the breakdown of glucose in the cytoplasm in respiration

Quadrat (3.2) square frame of known area used for sampling the abundance and distribution of slow or non-moving organisms

Radiation (3.6) energy in wave form such as light, sound, heat, X-rays, gamma rays

Receptor (1.4) cell surface protein which allows a cell to recognise specific substances

Receptor cell (2.2a) cell that can detect stimuli inside or outside the body

Receptor protein (2.2b) molecule which is complementary to a specific hormone molecule

Recessive (2.4) allele of a gene that only shows in the phenotype if the genotype is homozygous for that allele

Red blood cell (2.6) blood cell containing the pigment haemoglobin responsible for the transport of oxygen

Reflex (2.2a) a rapid action which protects the body from harm

Reflex arc (2.2a) pathway of information from a sensory neuron through an inter neuron directly to a motor neuron

Reliability (SSI) measure of confidence in the results of an experiment

Reliable (3.2) applies to results which can be trusted, or in which there is confidence

Replication (2.1) copying of DNA to produce chromatids before mitosis

Representative sample (3.2) a sample which is large enough and selected carefully enough to be truly proportional to the whole

Respiration (1.6) a series of enzyme-controlled reactions resulting in the production of ATP from the chemical energy in glucose

Respirometer (1.6) a device to measure respiration rates in living tissue

Ribosome (1.1) site of protein synthesis

Root epidermis (2.5) outer layer of cells of a root

Root hair cell (2.5) specialised cell that increases the surface area of the root epidermis to improve the uptake of water and minerals

Sample (3.2) a representative part of a larger quantity

Selection pressure (3.6) factor such as predation or disease that affects a population, resulting in the death of some individuals and survival of others

Selective advantage (3.6) an increased chance of survival for an organism because of possession of favourable characteristics

Selectively permeable (1.2)   refers to a membrane that controls the movement of certain molecules depending on their size

Self-renew (2.1)   property of stem cells by which they divide to produce more stem cells

Sensory neuron (2.2a)   nerve cell that transmits electrical impulses from a sense organ to the CNS

Sieve tube cell (2.5)   phloem cell without a nucleus and with sieve plates at the ends

Single gene inheritance (2.4)   discrete inheritance patterns produced by alleles of one gene

Small intestine (2.7)   part of the digestive system where most absorption of nutrients occurs

Source chromosome (1.5)   the chromosome from which the genetic material is obtained for transfer to another species

Source of error (3.2/SSI)   the origin of a mistake in drawing conclusions from experiments; a factor or feature of experimental procedure which can lead to mistakes

Specialised (2.1)   description of a cell that has become differentiated to carry out a particular function

Speciation (3.6)   formation of two or more species from an original ancestral species

Species (3.6)   organisms with similar characteristics and with the ability to interbreed to produce fertile offspring

Specific (1.4)   each different enzyme acts on one substrate only

Sperm cell (2.3)   gamete produced in the testes of male animals

Spinal cord (2.2a)   part of the central nervous system of a mammal that runs within its backbone

Spindle fibres (2.1)   threads produced during mitosis to pull chromatids apart

Spongy mesophyll (2.5)   plant leaf tissue with loosely packed cells and air spaces between them to allow gas exchange

Starch (3.3)   storage carbohydrate in plants

Stem cell (2.1)   unspecialised cell capable of dividing into cells that can develop into different cell types

Stimuli (2.2a)   changes in the environment detected by receptor cells that trigger a response in an organism (*sing.* stimulus)

Stomata (2.5)   tiny pores in the leaf epidermis that allow gas exchange (*sing.* stoma)

Stopwatch (SSI)   instrument for measuring time in an experiment

Structural (1.4)   referring to the proteins in membranes, muscle, bone, hair, nails etc.

Substrate (1.4)   the substance on which an enzyme works

Sugar (3.3)   energy-rich sub-units made by green plants in photosynthesis that can join into larger carbohydrates such as starch

Synapse (2.2a)   gap between two neurons in which chemicals transfer messages

Synthesis (1.4)   enzyme-controlled reaction in which simple substrates are built up into complex products

Syringe (SSI)   instrument for delivering small measured volumes of liquid

System (2.1)   a group of organs that work together to carry out a particular function, e.g. circulatory, respiratory and digestive

Target tissue (2.2b)   tissue with receptor protein molecules on its cell surfaces that are complementary to a specific hormone

Test tube (SSI)   glass container for holding liquid mixtures in experiments

Thermometer (SSI)   instrument for measuring temperature in an experimental situation

Tissue (2.1)   a group of similar cells carrying out the same function

Toxicity (3.5)   poison level

Transect (3.2)   a straight line of sample points across a habitat or part of a habitat, often marked by a long measuring tape

Transpiration (2.5)   evaporation of water through the stomata of leaves

Transpiration rate (2.5)   speed of water loss through leaves measured in volume of water per unit area per minute

Turgid (1.2)   description of a swollen plant cell with a full vacuole resulting from water intake due to osmosis

Ultrastructure (1.1)   fine structure and detail of a cell and its organelles revealed by an electron microscope

Unicellular (1.1)   single-celled

Vacuole (1.1)   membrane-bound sac containing cell sap in plant and fungal cells

Validity (SSI)   measure of the fairness of an experiment

Valve (2.6)   structure that prevents the backflow of blood within the heart and in veins

Variable (SSI)   factor in an experiment which is changeable or can change

Variation (2.4)   differences in characteristics that can be seen between individual members of a species

Vein (2.6)   general name for a blood vessel with valves that transports blood to the heart

Vein (2.5)   tissue consisting of xylem vessels and phloem tubes running through a leaf

Vena cava (2.6)   vein carrying deoxygenated blood to the heart from the body systems

Ventricles (2.6)   lower chambers of the heart that receive blood from the atria and pump it into arteries

Villi (2.7)   finger-like projections of the small intestine lining providing a large surface area for absorption of nutrients from blood (*sing.* villus)

Water bath (SSI)   trough of thermostatically heated water to provide steady controlled temperatures

White blood cells (2.6)   phagocytes and lymphocytes

Xylem vessels (2.5)   narrow, dead tubes with lignin in their walls for the transport of water and minerals in plants

Yeast (1.6)   unicellular fungus used commercially in the brewing and baking industries